MODULARIZATION

MODULARIZATION

THE FINE ART OF OFFSITE PREASSEMBLY
FOR CAPITAL PROJECTS

MICHAEL KLUCK AND JIN OUK CHOI

WILEY

Published by John Wiley & Sons, Inc., Hoboken, New Jersey
Published simultaneously in Canada

For general information on our other products and services or for technical support, please contact our Customer Care Department within the United States at (800) 762-2974, outside the United States at (317) 572-3993 or fax (317) 572-4002.

Wiley also publishes its books in a variety of electronic formats. Some content that appears in print may not be available in electronic formats. For more information about Wiley products, visit our web site at www.wiley.com.

Library of Congress Cataloging-in-Publication Data is applied for
Hardback ISBN: 9781119824718

Cover Design: Wiley
Cover Images: © Photograph by Michael Kluck

SKY10042309_020123

Michael Kluck:

This book is dedicated to my patient and loving wife, Jeanette, who endured the invasion of her home office space during the business world's Covid-19 transition from the office to the home—not only my day-to-day business activity and annoying online meetings that totally disrupted her schedule, but also all the late night and weekend hours when an idea or thought hit me and I felt the need to incorporate it immediately (for fear I might lose it. . .).

Jin Ouk Choi:

I dedicate this book to my wife, So Hyun Bae, my new baby, who was just born in Feb. 2022, Sooho Choi, and my parents, Bok-Gil Choi and Se Young An. There is no doubt in my mind that without their continual support and love, I could not have completed this book. Also, a special thanks to Yoon Jeong Choi and Kyle Shackleton for all their support and love.

CONTENTS

PREFACE

We would like to begin with a couple of remarks about the title of this book: *Modularization: The Fine Art of Offsite Preassembly for Capital Projects*.

What is in a title? We probably spent more time going back and forth on the title for this book than what was necessary. But it was the first thing you saw when you picked up the book, so a short explanation is probably due, assuming you even read this Preface. We wanted something catchy but, more importantly, descriptive and indicative of what the book is about.

Modularization is not new, and it is not revolutionary. It has been applied to industrial plants both onshore and offshore for over 50 years and incorporated into many major construction efforts. However, the problem has been that, in many cases, it was not correctly incorporated or implemented. More often than not, the result was a less than desirable project solution and a subsequent conclusion by the project team something along the lines of: "modularization was wrong for the project" or "modularization was a bad idea." This (wrong) conclusion was then carried (along with the resulting bad reputation of the module concept) into the planning of the next several projects the team tackles.

This is where the "art" in the title comes in and a quick side note about the two authors. We are engineers by background and probably are blessed with maybe only one or two artistic bones in our collective bodies. But, we both recognize a great work of art, the timelessness of a piece of classical music, and anything else truly great for that matter. We cannot tell you what makes these works great because, at face value, the inputs are the same—oils, brushes, and canvas for the painting or notes and words on a piece of paper for the musical score. So obviously, it is the way the tools and materials are combined and used.

A successful module project is quite similar. The modular project uses the same inputs as the stick-built project, but they must be "mixed up" in a complex combination of slightly different approaches and then implemented in a certain sequence to be successful. Unfortunately, this sequence is different in many aspects than that of the traditional stick-built construction project.

Fortunately, we do know and understand what makes a module job "great," and we want all of you to also understand what it takes. That is the aim of this book and why it was written. We are not so pompous or ostentatious to think the module planning identified in this book will ever be compared to the likes of a great work of art or a piece of classical music. But, after reading this book, we believe that you will have a better understanding of what it takes to make a module job successful to the point where you can successfully implement it on such a project.

So, it is hoped that the title and this short Preface stimulate your curiosity enough regarding this age-old alternative to the standard stick-built process for plant design and construction to continue reading.

Thanks for indulging us.
MPK & JOC

ACKNOWLEDGEMENTS

Michael Kluck

First, to my co-author, who allowed me to continue my quirky way of writing, all the while patiently taking it in and then suggesting adjustments that made it sound so much better! For his writing and input into all the chapters (more than he will ever take credit for). For his attention to detail. And, just as importantly, for his help in the organization of this effort that very quickly got bigger than both of us ever imagined.

Finally, to all the folks associated with all the module projects I have ever been involved with. I know many of you forgot more than I ever was able to pick up, but I am thankful for the interactions, the trust placed in me, and the lessons learned, some very painfully. Without all of you collectively, this work would not have the depth it needed to really explain "the fine art of pre-assembly."

Jin Ouk Choi

This book would not have been possible without the support from many individuals. I thank all the people involved in this book.

First, I really thank my co-author, Michael Kluck, who wrote this awesome book with me. Without Michael, this book really would not have been possible. I really enjoyed writing this book with him. With his great energy, passion, vision, and expertise on modularization, this book was possible.

Second, I appreciate the Project Management and Construction Engineering Lab (PMCEL) members' support. I would like to especially thank two PMCEL members: (1) Dr. Seungtaek Lee for supporting Chapter 3 Industry Status on Modularization, and (2) Dr. Binit Shrestha for supporting Chapter 13 Innovative Technologies for Modularization.

Finally, I would like to thank my loving and supportive wife, So Hyun Bae, and my sister, Yoon Jeong Choi, for their advice, encouragement, and support. In particular, Yoon's help and graphical advice on the figures were invaluable.

INTRODUCTION

Why is there an entire book devoted to the "art" of modularization? Simply stated, because the decision to modularize impacts every aspect of a project's planning and execution. Even from the very beginning of a project, in the Opportunity Framing (pre-FEL-1 or FEL-0) phase, some basic but important decisions must be made to avoid inadvertently limiting the modular opportunities.

So, the book has been written as a guide of sorts, walking through what is necessary to perform a modular project successfully, starting with some very basic "stuff." The path and sequencing of the book make perfect sense to us . . . to others, perhaps not so much. And while there may be a temptation to skip over these early parts to find answers needed in the later chapters, it is recommended that you ground yourself with the basics provided by these first few chapters. This will help set you up for the detailed module planning effort—just like a quick review of the rules of any game you plan to play is recommended prior to launching into it.

This book combines research findings from the analyses of real case modular projects with industry examples to help tie any theories presented to real-life scenarios. Thus, we expect that not all chapters will be of equal value to you. We also expect you may not even agree with some of the ideas and concepts stated. But we do hope that you can see past our methods on the approach and oversimplified examples to the reasoning and understand the conclusions from them.

So, consider this as a guide for your modular road trip. And, like a road trip, it's the journey, not just the destination. To be effective, such a guide needs to point one in the right direction/provide a preparation plan/suggest basic supplies/identify support along the way/and offer an emergency kit for when things don't go as planned. We hope the specific guidance in the format provided is helpful in implementing successful modular project planning, even if presented in a bit different format than you might have anticipated.

In order to help you better understand how the book has been written, the remainder of this introduction delves into a brief description of what each chapter is about, beginning with the basics of what modularization is.

Chapter 1: What Is Modularization?

Chapter 1 sets the basics in terms of what is needed from a ground-level or basic understanding of the concepts and terms. While this may seem too elementary, it is critical because everyone has their own definition of what a module is (or what it is not) and, as a result, a pre-conceived notion of what is involved in its development. It addresses the current state of industrial construction and why it continues to see productivity decline to the point where projects are becoming unsustainable with the traditional stick-built methodology. In addition, other sectors, such as manufacturing and shipbuilding, are examined in terms of how they have addressed their productivity issues and succeeded in a major turnaround in cost and schedule and how modularization offers a similar potential advance in both. Then, based on the foundation laid that "modularization is this industry's alternative," details are explored around the variability of the optimal modular answer and how a one size does not fit all. Finally, the chapter further grounds expectations by discussing "what modularization is not."

Chapter 2: Advantages and Challenges of Modularization

Chapter 2 follows the basics described in Chapter 1 with further details on the modularization concept, including a high-level look at why the module fabrication yard is such an ideal place for module assembly. More details are provided in terms of the advantages of selecting this project approach and some of the disadvantages (or challenges) of this project execution method.

Chapter 3: Industry Status on Modularization

Chapter 3 describes the industry status on modularization through a compilation of 25 actual case projects in terms of observed advantages, cost and schedule savings compared to the stick-built approach, difficulties, impediments, business case drivers for modularization, types of module units, module numbers, size, and weight, and characteristics of the job site and the modular fabrication shop, along with the subsequent editorial comments regarding some of the results themselves. This chapter validates the advantages and disadvantages which were initially identified in Chapter 2 via this set of actual project case histories and sets the stage for the following chapters, which address how to properly set up and execute a modular project in more detail.

Chapter 4: What Is a Module?

Previous chapters defined modularization and module terms, mentioned the advantages and challenges, and spoke about WHY the industry needs to really embrace this alternative to stick-built execution, but they did not provide details on the modules themselves. Chapter 4 explains the length and breadth (and height, no pun intended) of module variability. Modules are identified and described in terms of their more common types with the goal of providing a basis for understanding the magnitude and variability of options available when the term "modularization" is mentioned. Further, this chapter dives into the "module considerations" necessary to develop a successful module project concentrating on those that are slightly different and unique to the module philosophy, such as:

• Plot Plan development

• Differences between the stick-built and module layout

• Optimal module size

• Typical module contents and those that might require special considerations

• Guidance on division of responsibility in terms of material procurement.

Chapter 5: The Business Case for Modularization

Chapter 5 begins the discussion of the modularization business case, as a logical extension of the information provided by the previous four chapters that identified the basis of understanding what a module is, why modularization is important, why implementation will be a challenge, and the potential options in terms of module configuration. In addition, it explains what makes a project a good candidate for modularization, the factors to consider, the importance of timing in modularization considerations, the modular project execution planning steps by project phases, and describes how to conduct a business case analysis using the tool provided. It also provides guidance on how to approach the project module option in terms of developing the project-specific details necessary to identify the optimal module case for your specific project in terms of size, number, cost, schedule, etc. The three key contents addressed in this chapter are as follows:

• Important **Business Case Considerations** are identified and explained.

• **The 13-step Business Case Model** is explained in detail.

• A usable **Business Case Financial Analysis Model** is provided.

Chapter 6: The Module Team and Execution Plan Differences

This chapter presents how to manage the project philosophy shift to the modular approach via the Module Team, identifying key team members and their qualifications. The dynamics of the team are explained and contrasted against the typical stick-built project team. Examples are provided on team organization as well as how the team should be incorporated into the overall project management. Also, a deep dive is taken into explaining how the execution of the modular project differs from the stick-built project. These major execution plan differences are identified in terms of what they are, when they should first be incorporated into the project execution, and their priority for implementation within each project phase. This detailed listing is visually summarized in an activity table showing the various execution plan differences by project phase and the priority of execution within each project phase.

Chapter 7: Key Critical Success Factors for Modular Project Success

Chapter 7 complements Chapter 6 because, when considered together, the two chapters provide the foundational basis for why the modular job is so different and why understanding these differences and success factors are so important to the success of a modular job. Chapter 7 explains the concept of critical success factors (CSFs), how they were developed, and their importance to the success or failure of a modular project. The 21 CSFs are equated to a listing of common mistakes identified on modular projects that failed or were only marginally successful. Later, the chapter examines the CSFs in terms of difficulty of accomplishment, and identifies an ideal time for initial implementation and who has primary responsibility. Furthermore, it demonstrates the relationship between the CSFs and project performance. Finally, a somewhat whimsical exercise involving the Module "Perfect Storm" is provided, a hypothetical scenario of a module job where almost everything aligns in terms of the worst possible outcomes, and where we comment on the subsequent conclusions reached.

Chapter 8: The Fabrication Yard

Chapter 8 continues the explanation, initiated in Chapter 2, of the benefits a fabrication yard can provide in terms of the overall modular project execution. Specifics are explained in terms of location, physical size, and layout, and operational philosophy. Project guidelines for yard selection are suggested in terms of size, location, complexity, and number utilized. In addition, contracting strategies are discussed in terms of pros and cons as well as what seems to work best. Options on the division of responsibility regarding who provides what are addressed and suggestions provided.

Chapter 9: Module Considerations by Project Group

Chapter 9 takes everything provided in terms of basic "learning" in the first eight chapters and starts using this modular basis of understanding to look at and make decisions on how to approach some of the different day-to-day aspects of the typical module project in terms of the various groups involved. This chapter gets down to the "nuts and bolts" or, as some suggest, "gets into the weeds" of the modular analysis. It is where "the devil in the details" of the previous general modular statements is exposed. The chapter is filled with personal experiences—both good and not so good. But the content of these module-supporting details is universal and provides an idea of the types of decisions one will face. Subjects include:

- Engineering considerations: module evolution, timing, discipline leads, fab yard coordination

- Fabrication considerations: structural members, welding vs. bolting, sub-assemblies, weight control

- Completion/testing/pre-commissioning: what is "complete," pre-commissioning, tradeoffs

- Shipping considerations: basic logistics, motion analyses, grillage and sea-fastening

- Load-out: methods, design considerations

- Movement to site and hook-up: design considerations and the single weld hook-up (SWHU).

Chapter 10: A Practical Module Development Process

Chapter 10 is our "second most favorite chapter" (sic) of the book in terms of its development and the culmination of the previous nine chapters. Chapter 9 took all the basics from the previous chapters and used this information as a basis to help identify and resolve issues that typically come up when planning and implementing a module project from the technical standpoint. Chapter 10 takes a slightly different approach, still going all the way back to the left on the project timeline (to Opportunity Framing) but walking through potential interactions required during the early stages of project development. It also provides a detailed 5-step method for early module screening analysis that builds on the 13-step business case discussed in Chapter 5. Other specific items covered include:

• Module tenets: common misunderstandings in terms of the modular philosophy

• Initial project analysis requirements and timing of required discussions and decisions

• Recovery options should the module decision be made late in the project life.

Chapter 11: Modularization Application Case Study Exercise

Chapter 11 is the development of a composite "made for this book" modular project case study exercise. The hypothetical case study exercise walks the reader through a modular project development, beginning with a summary of the project and proposed facilities. Each phase of the project is developed with information typically available at that time in the project development. The reader must take the information available and, based on what is known, make assumptions and develop a plan forward, answering the questions at the end of the project phase. This effort is reiterated through each phase of the project—Opportunity Framing/ Assessment/Selection/Basic Design/Engineering, Procurement, and Construction (EPC). To keep the exercise on a consistent path, a primary solution has been developed that is used for "re-setting" the exercise for all, so there is a consistent approach at the beginning of each of the subsequent project phases. As a textbook, this solution is not part of the printed version but available for instructors (not for students) as a separate supplement. Contact Dr. Choi for the solution file: choi. jinouk@gmail.com

Chapter 12: Standardization: The Holy Grail of Pre-Assembly

Chapter 12 is the next step in the evolution of an effective implementation plan for the industrial capital project: Standardization. It starts with a brief explanation of the potential in cost and schedule savings, referencing both the historical gains of the shipbuilding industry as

well as the more recent CII UMM-01 research on standardization. It starts by describing the two different paths to integrating standardization with modularization and the benefits, tradeoffs, and basic economics of each. Later, it provides a road map in terms of identifying what makes a good candidate, when to start the evaluation process, how to approach the evaluation effort, and some critical success factors associated with the standardization concept.

Chapter 13: Innovative Technologies for Modularization

Modularization techniques have recently evolved thanks to the advances in and incorporation of certain innovative technologies. Chapter 13 examines the part that technology plays in modularization in terms of a few key emerging innovative technologies that can help the industry implement the modular technique more successfully on more complex and sophisticated projects. Those technologies of interest are:

• visualization, information modeling, and simulation

• sensing and data analytics for construction

• robotics and automation.

Chapter 14: Moving Forward

Chapter 14 is dedicated to a semi-formal wrap-up of the book along with some "thought-provoking ideas" on what the future of our industry might be. It reiterates the authors' goals and provides seven specific industry "Accelerators" that will make the future a more module-friendly world, if implemented. Those "Accelerators" are:

1. Applied Knowledge

2. Different Academic Teaching Approach

3. Identify, Acknowledge, and Incorporate Required Paradigm Shifts

4. Friendly Contracting

5. Industry Re-branding

6. More Alliances and Research

7. Planning Techniques and Their Combinations.

Chapter 15: Key Literature and Resources on Modularization

Chapter 15 provides a resource guide that can be used for further follow-up on the future actions suggested for our industry in Chapter 14. It lists reports, tools, and academic papers on modularization and standardization that may be considered beneficial in terms of a follow-up study or research—many of which were referenced at one time or the other in the preceding chapters.

chapter 1 What Is Modularization?

This chapter starts by answering the question, "What is modularization?" and setting the basics in terms of what is needed from a ground-level/basic understanding of the concepts and terms. Starting with a definition of terms, including modularization, module, and percentage modularization, it then introduces an industry best practice, "Planning for Modularization," and some basic modularization/pre-assembly philosophy. Next, it goes into the current state of industrial construction and discovers why it continues to see productivity decline to the point where projects are becoming unsustainable with the traditional stick-built methodology. A description is given of how other industries, including manufacturing and shipbuilding, have addressed their productivity issues and succeeded in a major turnaround in cost and schedule and how modularization offers a similar potential advance in both. Then, based on the foundation laid that "modularization is an alternative," details about the variability of the potential optimal modular projects and how one size does not fit all are explained. Depending on the project type and configuration, the analysis result can range from no reason to consider modularization to modularizing as much as possible. Finally, what is needed to get started with considering modularization as an option as well as "what modularization is not" are presented.

1.1 Definitions

1.1.1 Modularization

The authors define modularization as follows:

Modularization is the project business/execution strategy that involves the transfer of stick-build construction effort from the jobsite to one or more local or distant fabrication shops/yards, to exploit specific strategic advantages.

Very simply put, it is the fabrication, assembly, and testing of a portion of a plant or manufacturing facility away from its final site placement. The key is a conscious shift in project execution strategy.

Our definition is aligned with (Tatum, Vanegas and Williams, 1987) and (Construction Industry Institute, 2013; O'Connor, O'Brien, and Choi, 2013).

- Modularization is used to describe a process in which the principal construction method is the use of off-site prefabricated, totally preassembled and pre-finished modules (Tatum et al. 1987).

- Modularization entails the large-scale transfer of stick-build construction effort from the job site to one or more local or distant fabrication shops/yards, to exploit any strategic advantages. Thus, modularization may be considered a form of project business/execution strategy (Construction Industry Institute, 2013; O'Connor, O'Brien, and Choi, 2013).

Note that for the purposes of future reference and proper understanding, all pre-assembly efforts should be considered as part of the term modularization. This is a simplifying assumption on the part of the authors in order to eliminate the typical discussion that ensues regarding how to define and separate the two terms or differentiate between them. ANYTHING done to move work off site will fall under the term modularization.

Unfortunately, not all modularization (or pre-assembly) solutions end up being beneficial. The very fact that this book is being written is due to numerous examples of how a modularization effort was not properly implemented and ended up not adding value to a project. Since these efforts also fell under the term "modularization," the last clarifying phrase was included in the definition, ". . . *to exploit specific strategic advantages*." Another way we sometimes clarified this caveat was ". . . *that, when properly executed, provide best value for the project*."

So, what makes a project a modular project? Is it any amount of pre-assembly being performed or having at least one bona fide module in the mix? Is it exceeding a specific minimum percentage or a specific minimum ratio of modules to stick-build? Probably none of the above. As mentioned in the definition of modularization, it is a conscious shift in project execution strategy. So, a project becomes a modular project when the project team makes the specific conscious changes in job planning and execution in order to facilitate a modular outcome.

It would be nice if a magic wand could be waved over all the modularization projects, and those which were not implemented properly could be magically called something else. This would eliminate all the bad press, wrong conclusions, and in general erroneous implementation efforts that resulted in giving this term and this concept the bad rap that it currently carries. We have no such magic wand, so education in the proper approach to modularization and its planning is the next best solution.

1.1.2 Module

So, what is a module?

A module is a "portion of plant fully fabricated, assembled, and tested away from the final site placement, in so far as is practical."

(Construction Industry Institute, 2013; O'Connor, O'Brien, and Choi, 2013)

Again, the authors have taken some liberties in terms of this definition with some simplifying assumptions. The module can be very large or as small as a simple sub-assembly of a skid. It can be fabricated and assembled halfway around the world in a sophisticated fabrication yard, or it can be put together on a plot of land immediately adjacent to the actual project site. There are benefits with both scenarios, and we do not want to limit what can be considered in terms of options for the alternative to stick-building by unintentionally limiting the definition of these terms.

1.1.3 Percentage Modularization

As one starts using modularization on projects, one of the first questions management has is: "What percentage of your project is being modularized?" In many cases, the answer received is incorrectly used by management. In some cases, the assumption is that the higher the percentage of modularization, the more significant the schedule or cost savings a project has. In other cases, it may even be used to compare completely different projects by a similarly incorrect metric—the higher the percentage modularized, the more efficient the project execution.

While the above simplifications can occasionally be true, they frequently are not and instead become a slippery slope to start down. As we will explain in future chapters, there are just too many factors that influence what determines the "optimal" percentage modularization for a specific project, many of which have nothing to do with the actual module itself but depend on the site location, type of equipment, labor, and material costs, etc.

So, we will just simplify and say it straight up: To simply make a direct relationship between percentage modularization and project efficiency is wrong! There are just too many other factors that will impact the optimal solution. With the above preface, here is the definition:

Percentage modularization is the portion of original site-based work hours (excluding site preparation and demolition) exported to fabrication shops.
(Construction Industry Institute, 2013; O'Connor, O'Brien, and Choi, 2013)

We have attempted to qualify the definition to pertain to the actual erected plant or facility by removing variables associated with the differences in site location, these being the potential demolition of existing facilities or naturally occurring geographical impediments as well as the surface preparation required to make a site ready to physically support the proposed construction effort for the project. These efforts and costs are not to be ignored for the overall project economics, but it is important that such costs are not included in any metrics developed on modularization comparison as they could be so vastly different from one site to the next. With this basic understanding of a few of the terms that will be used throughout this book, we can now move forward on why modularization is important to our industry.

1.2 "Planning for Modularization" as a Best Practice

The Construction Industry Institute (CII) has identified 17 construction best practices (Construction Industry Institute (CII), 2021). By definition,

---

Best practices improve performance not only in terms of cost, schedule, and safety, but they also increase the consistency and predictability of project performance. By improving the consistency of project delivery, a company will have a better chance of improving project performance over the long term. This combined benefit of best practice use will likely give companies a distinct competitive advantage.
(Construction Industry Institute (CII), 2017)

Basically, to simplify this, **a best practice is a methodology that is critical to the success of a project.**

Below is a quote from the CII article publicizing the promotion of Planning for Modularization as CII's newest best practice back in 2015.

Planning for Modularization is not necessarily focused on promoting or marketing the concept of modularization, but rather on helping project teams better understand whether modularization is the right strategy for the project and, if so, how they can successfully implement the strategy to achieve improved outcomes in cost, schedule, safety, and quality while mitigating issues such as a lack of skilled craft labor or extreme weather.
(Construction Industry Institute, 2015)

Not only CII but many renowned agencies and institutes also highlighted the potential of modularization. Below are some quotes from them:

New trend, the reemergence of prefab & modularization tied to current influential construction trends, such as the increasing interest in lean construction, the rising use of BIM technologies and the growing influence of green construction.
(McGraw Hill Construction, 2011)

Modularization technology has evolved — more sophisticated and complex facilities are subject to its implementation.
(Modular Building Institute, 2010)

Over the next 20 years, its growing prevalence could significantly advance the productivity and competitiveness of the capital facilities sector of the U.S. construction industry.
(National Research Council, 2009)

Why do we bring up this best practice reference and other renowned institutes' quotes here in our introductory remarks in Chapter 1? CII and many others have identified (Planning for) Modularization as having the potential to dramatically improve project outcomes. But, this planning needs to be simple and straightforward enough so it can be implemented. The problem is that this implementation path is not often very clear and oftentimes is neglected or badly carried out—hence the basis for the book.

1.3 Current State of the Construction Industry

The construction industry is in trouble. Current productivity rates are unacceptable, and the industry now as a whole is unsustainable. Labor costs have increased while actual skilled workers' "time on tools" has decreased. Regulations (of all types) have further restricted how the average worker can function on the job.

Furthermore, the construction industry is losing its skilled workers. They are getting older, retiring, or with the increasingly cyclic nature of the business, getting weary of the uncertainty of employment and moving to other professions. Making it worse, the industry is also not attracting many new younger recruits. Part of the reason is that the construction industry, as currently viewed by many, is not considered "glamorous" or appealing to these new younger people coming of work age. With computer and tech jobs in the forefront and the overemphasis on a "college degree" as the only path to success, the typical construction industry job is often seen as a bit dirty, low-paid, difficult, and possibly even beneath their dignity. Trades, such as welding, pipe fitting,

and instrumentation, which still pay handsomely and offer increased mobility and salary, are no longer considered as noble a profession as they historically have been. So, some of the potential students of these trades shy away from considering these trades as acceptable alternatives to college. As a result, the construction industry continues to suffer in terms of recruiting and maintaining an adequate skilled workforce necessary to support these project construction efforts.

With the recent trend of more large-scale mega capital projects (see Figure 1.1), not only in the number of projects but also in the greater share of investments in terms of dollars, these new megaprojects come with increased emphasis on holding both cost and schedule goals. However, unfortunately, "98 percent of megaprojects suffer cost overruns of more than 30 percent; 77 percent are at least 40 percent late" (see Figure 1.2).

Owners continue to enforce these cost and schedule goals by shifting more and more of the cost and schedule risk to the Engineering, Procurement, and Construction (EPC) contractors, who in turn push it down to their sub-contractors. With increased competition for the

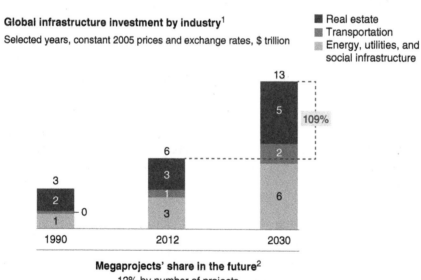

Global infrastructure investment by industry[1]
Selected years, constant 2005 prices and exchange rates, $ trillion

■ Real estate
■ Transportation
■ Energy, utilities, and social infrastructure

Megaprojects' share in the future[2]
12% by number of projects
77% by project value

[1]Forecast assumes price of capital goods increases at same rate as other goods and assumes no change in inventory.
[2]Project award date 2015 and beyond.

McKinsey&Company

Figure 1.1 Global infrastructure investment by industry. Source: McKinsey Productivity Sciences Center (2015). Reproduced with permission of McKinsey & Company.

Ninety-eight percent of megaprojects face cost overruns or delays.

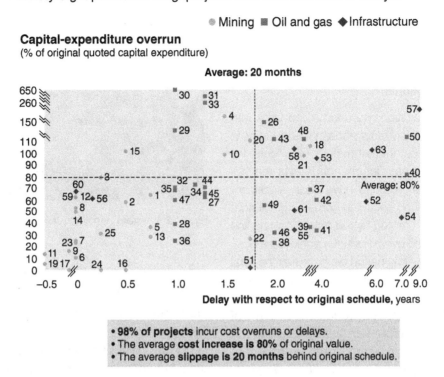

Capital-expenditure overrun
(% of original quoted capital expenditure)

● Mining ■ Oil and gas ◆ Infrastructure

- **98% of projects** incur cost overruns or delays.
- The average **cost increase is 80%** of original value.
- The average **slippage is 20 months** behind original schedule.

Figure 1.2 Megaprojects' capital-expenditure cost and schedule overrun. Source: McKinsey Productivity Sciences Center (2015). Reproduced with permission of McKinsey & Company.

fewer projects currently available in the market, the few successful EPC contractors must execute projects with razor-thin profit margins. Any deviation in labor pricing, material escalation, or schedule delays can spell financial disaster.

Modularization offers an alternative where the outcome is more predictable in terms of both cost and schedule, provided the required engineering and material inputs are provided in time. This is because the module fabrication yard operates more like a manufacturing facility

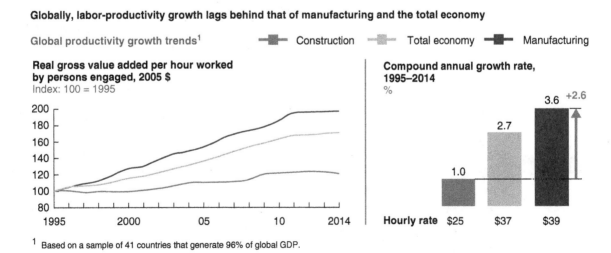

Globally, labor-productivity growth lags behind that of manufacturing and the total economy

Global productivity growth trends[1] ■ Construction ■ Total economy ■ Manufacturing

[1] Based on a sample of 41 countries that generate 96% of global GDP.

Figure 1.3 Global productivity growth trends. Source: McKinsey Global Institute (2017). Reproduced with permission of McKinsey & Company.

offering consistency in production. This will be further explained in Chapter 2 when we discuss the benefits and challenges.

1.3.1 How Did the Construction Industry End Up in Such a Sad State of Affairs?

To help answer this question, we need to go back into the past and look at the current trends in the industry, in particular the various sectors that make it up. Figure 1.3 identifies the issue that has not changed. Over the 20-year period shown above, there have been significant gains in the manufacturing sector, in some cases almost doubling productivity (as noted by the manufacturing line in Figure 1.3). In fact, the overall industry achieved a 50% increase over this same time period (as noted by the total economy line in Figure 1.3). However, during this same time frame, the construction sector at best was only marginally able to maintain constant productivity in terms of gross value added per hour worked (the construction line in Figure 1.3). While some may argue that the construction industry (Figure 1.4) has hit the ceiling in terms of how much of the typical stick-built efforts can be shipped off to shops for pre-assembly, there are major sub-sectors within the industry, e.g., automobile (Figure 1.5), or shipbuilding, that have shown significant increases in productivity, even rivaling the gains made by manufacturing.

So, what was the difference they managed to make when the construction industry failed? Here is a bit more detail on the gains achieved by the manufacturing industry, how they managed to do it, and what the construction industry is currently going through.

Figure 1.4 Construction industry. Source: travenian/iStock/Getty Images.

Figure 1.5 Automobile Industry. Source: xieyuliang/Shutterstock.com.

1.3.1.1 The Construction Industry Is Reluctant to Evolve

There are many reasons and factors that contribute to this downward trend, and an entirely separate study could be initiated to identify all the contributing factors, but for simplification purposes, in general, they all point to a reluctance by the construction industry to evolve. While manufacturing was embracing Lean and other processes that help optimize their efforts, and the shipbuilding industry was moving from stick-building to block-building to an Interim Product Database (in effect, a shopping catalog for interchangeable parts), the construction industry was still mired in the same production techniques used in the 1970s. These can be unflatteringly summarized as essentially gathering and moving a large number of skilled (or not so skilled) workers, who are unfamiliar with each other, to an unfamiliar construction site to work constructing a plant or process that many may not even be familiar with and perform a series of complex erection scenarios for those complicated plant processes.

In addition to the above, the construction workers have had to execute their jobs under circumstances that, as time progressed, require more and more requirements placed on them in terms of how they perform their work—additional training in terms of working together—traffic safety, isolation, lock out—tag out, working at heights, wearing special clothing for some tasks, gear for fall protection, etc. While most of these are positive in terms of safety results—reduction in near misses, medical treatments, lost time accidents (LTAs), and fatalities, they do take time to initiate, train, and implement. Add that to the current project unfamiliarity mentioned above, and this all adds to the reduction of the actual "time on tools" that the skilled worker needs to do the job.

1.3.1.2 The Waste Time in the Construction Industry

The manufacturing industry has effectively increased its percentage of the productive time versus waste, as shown in Figure 1.6, which is created using data from Diekmann *et al.* (2004).

According to Diekmann *et al.* (2004), only 43% of the entire active time is productive (including "Value Added" and "Support" activities/time, which means 57% of the work time is wasted or not adding values in the construction industry. At the same time, the manufacturing industry utilizes almost 74% of its work time productively. Of course, there has been significant improvement in both industries in recent decades, so the numbers are not the key point. The key message is that the construction industry needs to benchmark the manufacturing industry to improve its productivity. According to the same study (Diekmann *et al.*, 2004), the main reasons for the waste time in the construction industry were:

- Defects
- Overproduction
- Waiting
- Non-used resources

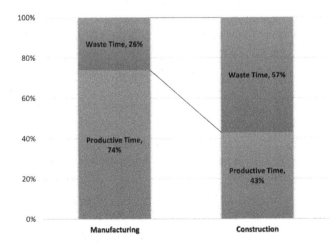

Figure 1.6 Waste comparison between manufacturing industry vs. construction industry. Source: Adapted using source data from Diekmann *et al.* (2004).

- Transportation
- Inventory
- Motion
- Excess processing.

In detail, there are diverse waste activities in the construction industry (Alarcon, 1997; Bhatla, Pradhan, and Choi, 2016):

1. Work not done
2. Unnecessary movement of materials
3. Re-work
4. Excessive vigilance
5. Unnecessary work
6. Extra supervision
7. Defects
8. Additional space
9. Stoppages
10. Delays in activities
11. Wastage of materials
12. Extra processing
13. Deterioration of materials
14. Clarifications
15. Unnecessary movement of labor
16. Abnormal wear and tear of equipment.

The sources of these wastes are (Alarcon, 1997; Bhatla, Pradhan, and Choi, 2016):

1. Unnecessary management requirement
2. Excessive control
3. Lack of control
4. Poor planning
5. Excessive red tape
6. Excessive quantity
7. Resource shortage
8. Misuse

9. Poor resource distribution

10. Poor quality

11. Resource availability

12. Theft

13. Unnecessary information

14. Defective information

15. Unclear information

16. Late information.

1.3.2 The Solution: Modularization

What if there was a way to identify all these "extra" wasted activities that limit the skilled worker s' "time on tools" and somehow get a construction team that has all this training, is familiar with their job, familiar with their setting and co-workers, and ready to go to work?

The answer, of course, is the modular solution—building parts of your plant in a module fabrication (in short, fab) yard. The module fab yard is typically located in a more temperate climate where weather is not extreme and where there is an ample skilled workforce and craft. The module fab yard creates an environment where the job moves to the worker. So, instead of having to deal with unfamiliar work surroundings, equipment, and team members, the worker is centrally located, and the jobs come to him/her in terms of "modules." This provides a synergy in which the major variability is reduced to the actual work itself—the module. Because the workers have a more or less steady workload, they can concentrate on honing their skills and becoming more proficient with their tools. The steady workload also creates a more defined split between work and personal life as well as an ability for the worker to maintain a balance between both—something that results in a more well-rounded and productive individual.

So, essentially, the more construction work hours we can move from a uniquely special project site to a more standardized fabrication facility, the potentially better outcome you can have in terms of efficiency and speed. In conjunction with improved technologies, the project inefficiencies are driving the shift to prefabrication and modularization.

This is the main reason for the current interest in the modular approach to project planning and execution.

1.3.2.1 Advanced Scheduling and Work Planning Techniques

The emergence of Advanced Work Packaging (AWP) from obscurity to an almost household phrase has revitalized, standardized, and propelled the age-old but good work practices of detailed solid project planning. Lean techniques have always been used by fab yards with high production assembly line fabrication.

Advance Work Packaging (AWP) All the various terms within AWP (Work Face Planning/Engineering—Construction—Installation—Procurement Work Packages) help one visualize the need for coordination throughout the project planning phases. This coordination to support the Path of Construction along with a continued confirmation that equipment and materials are on site prior to initiation of activities, are the basics of good project execution. Agencies like CII and others have been instrumental in providing both the research as well as the practical implementation guidance to help the construction industry understand how to properly plan. So, taking these basic project planning concepts and calling them something different (e.g., AWP) is a good thing if it helps bring better project planning to the forefront, even if the basic concepts are not new and have been practiced by some for a long time.

The module fab yard has always used good assembly planning techniques (the planning currently referred to as AWP). Their goal is similar to the assembly plants of the Ford Motor Company except perhaps on a grander scale. That goal is to get the maximum throughput (in terms of completed module tonnage) possible.

Lean Lean techniques have been used in manufacturing processes for many years. But they are also used in the module fabrication yard, even if they have not historically been labeled as part of a Lean construction or planning technique. For example, pipe shops internal to the fab yard are in the business to optimize the subassembly work of producing pipe spools. As such, they continue to optimize their approach to spool fabrication in terms of steps saved and just-in-time material and assembly techniques. The spools are fabricated under a schedule that considers both the optimization of size and metallurgy within the assembly line as well as their eventual need for installation to come up with the most efficient plan and sequencing to get the fabrication completed. This is the essence of many of the Lean techniques used in manufacturing.

1.3.3 Why Aren't All Jobs Modular?

All the industries mentioned above have a history that points to the need for a basic transformation in our current stick-built philosophy for industrial projects in order to get the substantial gains desired. So, why aren't we seeing a major shift in the philosophical approach to many current industrial projects?

The biggest reason for the reluctance to change is the fact that the stick-built approach to project execution is the one that is most familiar to the majority of the EPC companies and many of the client organizations. So, despite the trend of overruns on major EPC projects, the stick-built execution plan is the one that is typically chosen. Part of this is because of the contracting strategy by owners/clients. The stick-built approach allows for more decisions on detailed design to be shifted to the right (later in time), allowing these decisions to be made later in the project timeline. This fits well within an owner/client program where early funding may be an issue (e.g., it cannot be released until a financial investment decision, FID).

The stick-built method also allows for later engineering, later changes, and more flexibility on the job site. This bit of additional flexibility is welcomed by the EPC contractor, who may be contractually obligated to pay adverse financial penalties for late completion. The modular solution, in some cases, can put the contractor in an especially vulnerable position as its successful delivery depends on the accelerated design and material deliveries to the fab yard. This creates more complexity and risk that typically are not contractually shared with the owner/client. Unfortunately, this type of later planning also pushes a much higher percentage of the work to the construction team on the site to complete—a more expensive solution in terms of the total project cost.

1.3.4 Are Module Jobs More Expensive?

Second, at face value, in terms of a simple comparison between the stick-built method and the modular method (and without an exhaustive review and evaluation of the potential benefits), the module job may also *look like* it might do the following:

- **Cost more:** extra materials in terms of steel costs, logistics of final module shipment, and costs to hook up the modules at site.

- **Take longer:** you have the additional shipping duration from fab yard to the job site, as well as time to set in place and hook up.

- **Require additional and earlier engineering support:** the module "box" must be designed early, extra transport analysis is required, and follow-up at the project site for hook-up is needed.

- **Demand additional management at the fab yard:** in terms of a "second" construction management team (CMT)—while this may reduce the size of the site team, it is not a complete one-for-one reduction.

- **Need extra transport or shipping:** extra consideration in terms of vessel identification, transport loads on the structure, load-out, grillage and sea-fastening design for the ocean or land transportation, and, finally, special design for the setting of the final modules.

In fact, we will go further "out on a limb" to make a point and state that IF:

- you were able to provide all the required and qualified craft personnel exactly when needed (and they were good at their jobs with no quality issues needing re-work), AND

- the craft personnel were not only well versed in their area of expertise but also familiar with how their work related to the other crafts to the point that they worked as a single team, AND

- you were able to provide craft supervision with the best project construction management, AND

- you have all equipment and materials arrived just in time for installation by construction equipment that was also mobilized and set up only when needed, AND

- you effectively utilize this equipment because all the project equipment was on the site and the piping prefabricated and ready to install (and then quickly de-mobilize), AND

- you have all your craft and supervision not only located and living just off the site but completely happy with their current wages and benefits and willing to work at 100% until the last day, AND

- you have a local community in the vicinity of the project site that had no problem with all the temporary inconveniences they experienced due to the extra construction activity, AND

- all permitting posed no issues in terms of delays and problems, AND

- weather and seasonal changes played no part in any delays and shutdowns; in fact, you have no weather-related delays.

Then, modularization would indeed probably take longer, be more expensive, and not be for your project! However, coming to this conclusion is exactly the issue—we would again venture "out on a limb" here and also state that no project has ever had *ALL of the above luxuries present in the context of a project*. Of course, modularization does not provide all of them either, but in most cases, the module fab yard comes much closer to meeting most of them than the typical construction site they will be delivering to.

While many of the above-perceived disadvantages of the modular solution may seem to be insurmountable negatives, there are simple mitigation efforts that, when considered early enough, will more than offset all of the above negatives. As a teaser and while not wanting to give away the punch line of Chapters 2 and 5, one of the major modular benefits is the reduced cost primarily driven by the large difference in productivity and labor cost between the module yard and the proposed project site. This delta cost difference not only offsets the above "negatives" but also provides substantial savings in both cost and schedule. The detailed benefits and challenges of the modular solution will be discussed in much more detail in Chapter 2, Advantages and Disadvantages of Modularization, and Chapter 5, The Business Case for Modularization.

1.4 Three Distinct Module Options (or Circumstances)

While there are definite advantages to the use of modularization on all projects, the majority of the applications fall into three distinct options or circumstances (see Figure 1.7):

Figure 1.7 Three distinct module options or circumstances.

- very limited or no reason to consider;

- selective implementation through business case analysis;

- maximizing the full use of the module/pre-assembly concept.

1.4.1 Very Limited

The first category pertains to the type of projects that involve large earthworks, excavations, and/or in-situ concrete placements or pours. These types of projects primarily involve the movement or placement of large volumes of material—typically earth or concrete. Examples include large earthen works such as dams and retaining walls, in-ground concrete storage tanks, and water treatment facilities like the one in Figure 1.8. This does not mean that some of the supporting mechanical and structural parts, such as central piping supports and distribution arms, should not be considered, but the majority of the project's efforts will be associated with the excavation and installation of structures that are not conducive to being built off the site.

1.4.2 Selective Implementation

This option is by far the most likely to be implemented, probably because it covers the widest range of possibilities. This option runs the gamut from a single small pipe rack module to anything short of modularizing to the

Figure 1.8 Water treatment facility—very limited module opportunity. Source: Chalabala/iStock/Getty Images.

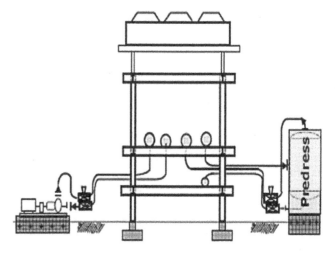

Figure 1.9 Typical stick-built approach.

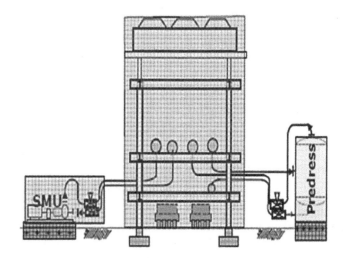

Figure 1.10 The same plant, but with pipe racks and skid-mounted units built offsite.

Figure 1.11 Photo example of same completed pipe rack. Source: Michael Kluck and Jin Ouk Choi

maximum extent possible. This option is another of the reasons for writing this book. Selective implementation is the "art" of figuring out the best value for your project money in terms of offsite fabrication. As we will cover in the subsequent chapters, identifying the optimal module solution is not a simple flip of a coin or a 5-minute exercise. Finding the optimal solution depends on a determination on a multitude of interacting decisions and project-specific criteria. Yes, this sounds very vague and nebulous, but we will explain what is involved in identifying the optimal solution over the next few chapters. If we decided to explain how to do it here, we would end up with a Chapter 1 that might be over 100 pages long. Instead, we ask that you be patient with us for a bit as we get the basics of modularization established. Once completed, we will delve into this option in much more detail in Chapter 5, The Business Case for Modularization.

In terms of a typical process plant example, below are two versions of selective implementation. The first one (see Figure 1.9) is a minimal approach in terms of offsite prefabrication where everything is stick-built except some of the vertical vessels, which are "predressed" with insulation, heat tracing if applicable, and some instrumentation.

The second one (see Figures 1.10 and 1.11) is the same process plant except that the pipe rack has also been considered as well as some of the equipment that has been packaged as a skid-mounted unit (SMU), with both designed and prefabricated offsite for shipment to the project site.

Neither option can be considered always right or always wrong. It depends on the project-specific factors that drove the project management to the decision.

1.4.3 Maximized Modularization

In some cases, the project's best bet (i.e., optimal solution) is to maximize the modularization that can be accomplished. If you think about it a bit, you can easily come up with what these projects might look like, for example, building a processing unit in the Arctic (or, for that matter, in the middle of some reclaimed mangrove swamp in Nigeria). Both extremes require maximization of this option due to the extreme hardship (or cost) of mobilizing a workforce of sufficient size and quality and continuing to support them in terms of accommodations (housing and nourishment), tolerable working conditions, personal and job-related equipment, recreational alternatives, and of course financial incentives for the duration of the project.

Using the same example project plant cross-section, the maximized modularization effort becomes one that not only takes the vessel and pre-dresses it but includes the pipe rack and piping as well as the equipment, all installed on a common skid and completely piped up, instrumented, and in some cases wired for power (see Figures 1.12 and 1.13). Such a configuration has taken the maximum labor required at the site "off site" and sent it to a fabrication yard, leaving the hook-up of adjacent modules at the site as the only remaining labor requirement.

Figure. 1.12 Same plant—maximum modularization.

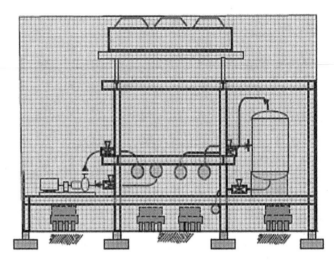

Figure 1.13 Example of maximized modularization.

But you do not have to build your project in the Arctic Circle or the middle of the Sahara Desert to have this as a potential optimal module solution. As we will further explain in Chapter 5, there may be specific areas of even relatively "civilized" parts of the USA where, due to a combination of one or more of the following, modularization is the better solution (O'Connor, O'Brien, and Choi, 2013):

- a lack of specific skilled crafts in a locale;

- a lack of infrastructure (for temporary accommodations);

- unresolvable issues with obtaining enough nearby temporary laydown area for the on-site construction support;

- problems with environmental restrictions (e.g., too much ambient light or noise, traffic, etc.);

- physical restrictions to and from project site due simply to baseline traffic loads on existing highways;

- the owner wants the project As Soon As Possible (ASAP).

For example, it does not take much in terms of losing an hour or so on both ends of a day due to a long commute along with having to staff the project site with inadequate quality and quantity of critical workforce to tip the economic scales in favor of an offsite fabricated project.

Again, as mentioned above on the selective implementation, there is still a point even in this maximum modularization option where you can modularize too much. We will go into details of this in Chapter 4, where we get into the philosophy of how to size a module, what goes into one, and what is best left out of one and shipped directly to the site.

1.5 What Modularization Is "Not"

Anyone keeping score in terms of "pro-module" (offsite fabrication) versus "anti-module" (stick-built) through the Introduction and up to this point in the first chapter will have the pro-module team totally annihilating the anti-module team to the point where some may begin to wonder:

• Why even have a book on this—just go do it!

Or

• If modularization is such a great thing, why is it not being used on every job?

While we discussed this briefly earlier in this chapter, more detail still needs to be provided to you in order to help put the module decision-making process into perspective. First, the traditional stick-built method of industrial project execution is still a very solid method to complete the "C" in EPC, as long as your project can get the necessary number and quality of skilled workers to the project site and keep them there until the job is completed and you have a management team that is excellent at working together and working to a completion plan. Second, the traditional stick-built method of industrial project execution is still very well-known and understood in many areas of the world, and it is effective as long as there are foremen, supervisors, project leads, and leaders involved who can effectively do exactly that—"lead." Third, there are certain things that modularization is not: "Panacea," "Constant," and "Easy."

1.5.1 Panacea (a Solution for All Difficulties)

Modularization is not a panacea. Modularization will not overcome a poorly planned and/or poorly executed EPC project. It will not overcome an overpriced but under-bid project. Furthermore, it will not take a mismanaged scheduling effort and put it back on schedule. Modularization is not a miracle cure. In fact, there are many more examples of poorly run projects that, when incorrectly implementing

modularization, managed to drive the project into realizing even more dismal results in terms of project costs and schedule than there have been results of those same projects benefitting from the positives of the poorly developed modularization decision and execution plan. This is because modularization is not a switch you turn on from off and everything suddenly happens. (Well, if that approach is taken, everything does indeed happen, but typically it just continues to "happen" to get much worse.)

Instead, the decision to modularize needs to be made early in the project life and with much forethought. One does not go into a project with a goal to modularize without a dedicated analysis driving that conclusion. Furthermore, what happens when this proper analysis approach is taken with respect to how the modular decision was made, the project team may uncover by this analysis that the optimal module decision is very different from that originally thought or suggested, for any one or several important reasons.

As mentioned earlier, extra efforts in various areas of a project need to be identified and executed to obtain the desired modular result. While some of these extra efforts are common with all projects, others will vary in importance depending on project specifics. But in all cases, as these extra efforts are either missed or ignored, the resulting opportunities for a beneficial module result diminish or may even disappear completely. Again, this may not provide the detail being sought, but much more detail will be provided in Chapter 7, Critical Success Factors for Modular Project Success.

1.5.2 Constant

The business drivers for modularization are not constant. They vary from project to project. For example, we previously talked about the obvious module efforts required for the areas of the world dominated by extremes in temperature and lack of infrastructure. These make for an easy decision.

The difficult areas are where most projects occur—in temperate climates with some sort of craft availability and established infrastructure for access. In these areas, a closer look must be provided, and the pros and cons of offsite assembly examined in detail to identify the optimal approach. Again, this will be discussed in much more detail in Chapter 5, The Business Case for Modularization, where we will go into how to approach the modularization decision in the more industrialized and populated areas of this world.

1.5.3 Easy

Modularization is not "easy." What we mean is that it is not a decision or conclusion that is made quickly and without forethought. We are not indicating that this is a "hard" decision to make or that this is an "expensive" or "time-consuming" decision process. In fact, if this discussion about the possibility of an alternative to on-site stick-building were brought up in the very early phases of a project concept (e.g., Opportunity Framing or FEL 0 or pre-pre-FEED), it can set the stage for opening up the project to the module alternative with little extra effort or cost. However, such a discussion must occur early enough.

For example, a major company has decided to build several processing plants throughout the United States. The details on the number of plants and strategic locations are still being kicked around. Alternative site locations should be evaluated in terms of availability to a major roadway or navigable water access. Even if specific details of the process are not defined, there typically are high cost and/or schedule savings if the process can be built in large enough "blocks" to allow a significant level of pre-commissioning to be performed prior to shipment. If this option becomes one of the factors in the very early analysis of the multiple plant program, then more intelligent decisions can be made in terms of potential project sites. More of this type of discussion will be covered in Chapter 11, Modularization Application Case Study, where we discuss Standardization (the next step).

1.6 Summary

Chapter 1 touched on many parts of the philosophy of modularization, why the construction industry needs it, how it is beneficial, what potential issues there might be, how it might be improved, and more importantly, why it is not universally implemented. Hopefully, it has provided the nebulous term (modularization) some meaning and dimension and set some limits on what can be expected

when pursuing this alternative project philosophy. The following chapters will cover all the above in more detail now that we have hopefully developed a common playing field with respect to what the terms are, what they mean, and how we are applying them.

References

Alarcon, L.F. (1997) Tools for the Identification and Reduction of Wastes in Construction Projects. In Alarcon, L. F. (Ed.) *Lean Construction*. Rotterdam: Balkema, pp. 365–377.

Bhatla, A., Pradhan, B. and Choi, J.O. (2016) Identifying Wastes in Construction Process and Implementing the Last Planner System in India. *Journal of Construction Engineering and Project Management*, 6(1), pp. 11–19. doi: 10.6106/JCEPM.2016.6.1.011.

Construction Industry Institute (2013) *Industrial Modularization: Five Solution Elements*. Austin, TX: The University of Texas at Austin: Construction Industry Institute.

Construction Industry Institute (2015) Planning for Modularization Becomes Next CII Best Practice. Available at: https://www.construction-institute.org/blog/2015/planning-for-modularization-becomes-next-cii-best (Accessed: 31 December 2021).

Construction Industry Institute (2017) *SP166-4 –CII Best Practices Handbook*. Austin, TX: The University of Texas at Austin: Construction Industry Institute.

Construction Industry Institute (2021) Best Practices. Available at: https://www.construction-institute.org/resources/knowledgebase/best-practices (Accessed: 31 December 2021).

Diekmann, J.E. *et al.* (2004) *Application of Lean Manufacturing Principles to Construction - RR191-11*. Austin, TX: The University of Texas at Austin: Construction Industry Institute.

McGraw Hill Construction (2011) *Prefabrication and Modularization: Increasing Productivity in the Construction Industry*. SmartMarket *Report*.

McKinsey Global Institute (2017) Reinventing Construction: A Route to Higher Productivity. Available at: https://www.mckinsey.com/~/media/McKinsey/Industries/Capital Projects and Infrastructure/Our Insights/Reinventing construction through a productivity revolution/MGI-Reinventing-Construction-Executive-summary.ashx.

McKinsey Productivity Sciences Center (2015) The Construction Productivity Imperative. Available at: https://

www.mckinsey.com/business-functions/operations/
our-insights/the-construction-productivity-imperative.

Modular Building Institute (MBI) (2010) *Improving Construction
Efficiency & Productivity with Modular Construction*.
Reston, VA: Modular Building Institute.

National Research Council (NRC) (2009) Advancing the
Competitiveness and Efficiency of the U.S. Construction
Industry. Edited by N. R. C. *Committee on Advancing the
Competitiveness and Productivity of the U.S. Construction
Industry*. Washington, D.C.: The National Academies
Press, p. 122.

O'Connor, J.T., O'Brien, W.J. and Choi, J.O. (2013) *Industrial
Modularization: How to Optimize; How to Maximize*. Austin,
TX: The University of Texas at Austin: Construction Industry
Institute.

Tatum, C.B., Vanegas, J.A. and Williams, J.M. (1987)
*Constructability Improvement Using Prefabrication,
Preassembly, and Modularization*. Austin, TX: The University
of Texas at Austin: Construction Industry Institute. Available
at: https://www.construction-institute.org/resources/
knowledgebase/knowledge-areas/design-planning-
optimization/topics/rt-003/pubs/sd-25.

chapter 2 Advantages and Challenges of Modularization

In this chapter, we go into more detail about the advantages and disadvantages of modularization. First, we look at the key characteristics of fabrication yards to better understand why they can bring advantages. And then, we go over the major advantages and challenges (or disadvantages) that should be included in the overall analysis of the module potential. In Chapter 3, we will describe the industry status of modularization, which will cover the advantages and challenges of modularization, again in detail with supporting case projects.

2.1 Why Do Fabrication Yards Have an Advantage?

2.1.1 Obvious Reasons

The obvious reasons are listed below and are the ones you would typically be presenting when trying to sell a modular approach.

- **Temperate climate:** Most fabrication facilities are in areas of the world that boast a long construction window in terms of weather. And even in the areas where weather may be a factor for part of the year, some have further added large fabrication buildings where segments or, in some cases, entire modules can be built indoors.

- **Building at grade:** The highest production efficiency is achieved when building on the ground (at grade) by a team who are familiar with the overall program as well as each of the other crafts and activities, so they work as a cohesive unit. This is why so much emphasis is placed on this.

- **Supporting lift equipment:** Because as much as possible is built at grade, all fab facilities come complete with lift equipment—multiple gantry cranes, crawler cranes, or jacking systems for lifting and setting upper decks on lower ones. These are set up in the assembly area to complement the specific method of assembly that the fab facility has chosen as its most efficient way to assemble.

- **Concurrent/parallel fabrication and construction:** Multiple modules can be fabricated concurrently in one or multiple fabrication yards, but modules can also be fabricated concurrently with the site development. This feature really helps the project to reduce the overall project duration.

2.1.2 Less Obvious Reasons

- **Captive craft labor:** Most competitive fabrication facilities/shops/yards are located near a large population area where craft have a relatively short commute to work. The fab facility provides these craft with a consistent and relatively constant workload, providing a reasonable work week and a sense of security that is usually not found in the construction industry that must travel from project to project.

While there is always a concern raised about a construction site near a fab facility actually enticing workers away from the fab facility, the fab facility typically has no problem recruiting and maintaining qualified craft. Many skilled laborers in the construction industry are more than willing to give up a small percentage of the wages that they might be able to make at a greenfield construction site in order to be guaranteed a standard workweek, benefits, a safe place to hone their skills, less commute time (and more time with family), and enough work to be challenging.

- **Skilled craft:** As mentioned above, because the fab yard has a constant influx of work, the craft has more opportunities to become "skilled" in a particular part of their craft. For example, stainless-steel welders can excel in alloy welding because there will always be enough alloy pipes to practice their craft on.

- **Teamwork:** Because the fab yard uses a standard manufacturing type process for the pipe spools and structural steel, which individually feed into a slightly different but also standard assembly line process for the modules, the relatively constant "core" craft workgroup and management know and understand the process of building the modules. As such, they can anticipate and offer support, working more as a team, each craft knowing what is expected of them and communicating what they expect to others.

- **Safety culture:** Most fab facilities have a combination of a manufacturing process—structural steel into plate girders, columns and beams and pipe and fittings into spools, as well as the assembly process of taking and moving these components into sub-areas that are further consolidated into decks or pancakes, to be stacked upon completion to create the multi-level module. Both the manufacturing and the assembly processes are repetitive enough to be conducive to the application of techniques of continuous improvements, such as Lean. These are typically part of a larger subculture in terms of the desire to improve the entire fab facility organization. The successful results are clearly visible when visiting some of these large Far Eastern fabricators where the desire to improve quality and safety procedures actually spills over into the rest of the activities, from personal tool care to area housekeeping, resulting in notable reductions in safety incident rates.

There is more to these reasons/advantages that needs to be brought out to further help in understanding them. Thus, we will go over some detailed basics about the "generic" fabrication yard separately later, including the benefits and characteristics, selecting the right fabrication yard, contracting strategy, and division of responsibility in Chapter 8 The Fabrication Yard.

2.2 Advantages of Modularization

There are many advantages (or benefits) of implementing modularization. The advantages are:

1. Reduced capital costs

2. Improved schedule performance

3. Increased productivity

4. Improved predictability (surety/reliability) or less variability

5. Increased safety performance

6. Increased quality performance

7. Increased sustainability performance (green benefits)

8. Site and site construction team benefits:

    a. Longer duration for foundations and underground

    b. Higher efficiency at the site

    c. Completion of pre-commissioning efforts off the sit

    d. Less impact on the site.

Note that most of these advantages are due to the abovementioned reasons. Now, we will explain these advantages in detail one by one.

2.2.1 Reduced Capital Costs

This is probably the number 1 reason for the application of modularization, followed quickly by questions on improving the schedule. These two are the most frequently cited reasons for this alternate form of on-site stick-built approach. Typically, the conversation with an owner/client begins with how modularization of a specific area will be able to cut costs. This is often brought up where some of the project sites have limited craft availability, limited construction laydown areas, limited access to the site for a large number of employees, or some other combination where moving total craft work-hours off the site will make sense.

2.2.1.1 How Does Modularization Reduce Cost?

This topic is explained in much more detail in Chapter 5, The Business Case for Modularization, but the main driver is the comparison of the difference in the Relative Work-hour Costs (RWC; labor productivity factor times

Description	Project Site	Selected Fab Yard
Labor Productivity Factor (Hours required to complete one unit of work; based on Ideal USGC = 1.0)	2.5	1.3
All In Wage Rate (AIWR) (local cost / work-hour)	$135	$45
Relative Work-hour Cost (RWC) (cost / hour for a similar unit of work)	$337.50	$58.50
		Project Site to Fab Yard
	Economic Productivity Ratio	5.77

Figure 2.1 Economic productivity ratio example.

craft labor cost ($/work-hour; typically quoted as All In Wage Rate [AIWR]) of the fabrication yard to that of the project construction site. In all cases, the RWC at the fabrication yard will be cheaper than the same work-hour at the construction site, or an Economic Productivity Ratio (EPR) of the project site to fab yard higher than 1. A simple example (see Figure 2.1) describes the RWC and the EPR.

In Figure 2.1, the project site has a lower labor productivity factor (2.5) than the fabrication yard (1.3, with 1.0 as the US Gulf Coast site (USGC) baseline). Note: Anything lower is more efficient while anything higher is less efficient. At the same time, the cost of this craft labor (AIWR) at the project site ($135/work-hour) is much higher because, in this example, the construction site is located in a remote site where labor is scarce while the fabrication yard is located in SE Asia, where the cost of labor ($45/work-hour) is much cheaper. The RWC at the project site is $337.50 (= 2.5 x $135.00), which is higher than the RWC at the fab yard, which is $58.50 (1.3 x $45.00). As a result, the EPR of the project site to the fab yard is 5.77 (= $337.50/$58.50). This means the module yard is more than 5.77 times as economically productive as the project site. Putting it another way, every craft work-hour taken off the project site will support over 5.77 craft work-hours at the fabrication yard. As a result, the module fabrication yard will expend only 17% (= 1.0/5.77) of the estimated site cost (for an equivalent scope). With this large difference, this project should strive to maximize the reduction of work-hours at the site by moving them to the module fabrication yard.

2.2.2 Improved Schedule Performance

And, like a one-two punch, as soon as cost benefits are provided, the owner/client immediately asks what the schedule improvement will be. Figure 2.2 shows the typical schedules of a stick-built project and its corresponding modular option.

As you can see from Figure 2.2, which is the comparison between stick-built project schedule versus modular project schedule, a modular project can cut total project duration significantly compared to the traditional stick-built project. This schedule reduction is due in part to the ability to work concurrently at the construction site as well as at the module fabrication yard. The engineering efforts, in either case, are about the same in terms of duration between the stick-built and module project schedules, but some of the deliverables are internally adjusted to reflect support to the fabrication yard. This allows the project to either mobilize later at the site or mobilize as they would with the stick-built effort and take longer to get out of the ground in terms of the underground piping, electrical, and foundation efforts.

This schedule improvement can provide an opportunity for the owner/client to reduce turn-around time for operating facilities, deliver products faster to the market, and reduce financing liabilities. Also, the contractor can benefit from this schedule improvement as it reduces the project cycle times, financing liabilities, and work in progress.

2.2.2.1 Cost or Schedule Benefits?

So, with any analysis of benefits, these cost and schedule benefits go hand in hand during early discussions and ought to be arranged together. As noted above, this advantage quickly follows cost savings as a discussion point, so an early discussion must be initiated with respect to which one has priority. The client will be quick to say "Both!" This is not an acceptable answer because a cost-driven module program could be different in its development than a schedule-driven module program. Since this is sometimes difficult for an owner/client to understand, we like to pose the following question regarding priorities: "Is it more important that you save $50 million or reduce schedule by 2 months?" (use whatever numbers fit the specific project).

This line of questioning will hopefully force the owner/client to go back to their project economics and further evaluate the Net Present Value (NPV). The owner/client knows the project particulars with respect to the profit and cash flow details that they may not wish to share with the EPC contractor. However, the module implementation effort varies between the two drivers, so

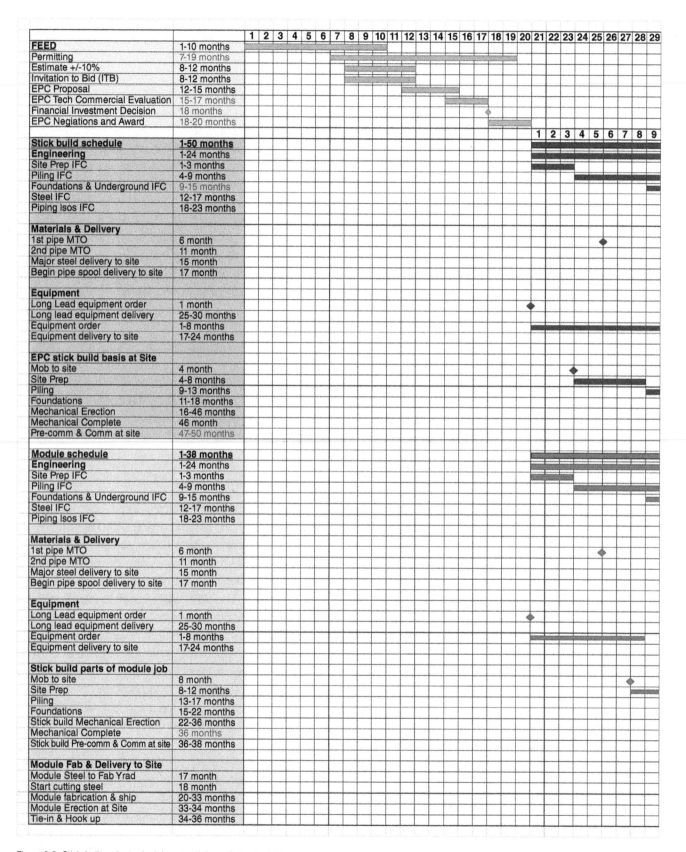

Figure 2.2 Stick-built project schedule vs. modular project schedule.

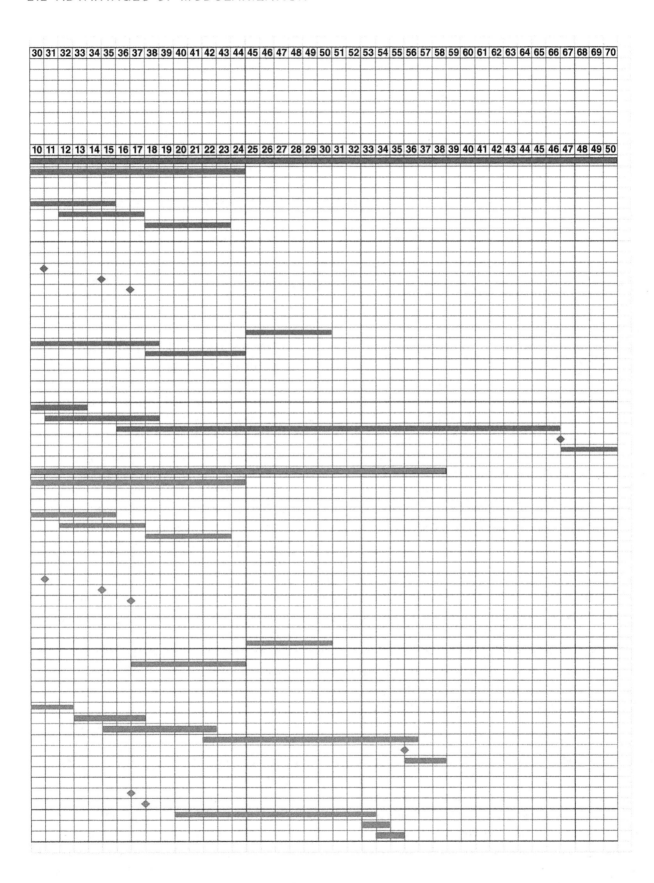

it is important that the client performs this analysis and provides the results in terms of which one should take priority, so the EPC contractor can make the appropriate adjustments in terms of driver priority. The owner/client knows which one has a higher priority and can come to a decision on it but will need additional input from the EPC contractor on why. So, it is up to the EPC contractor proposing modularization to educate the owner/client on why this priority is vital to the project outcome. More discussion of this in Chapter 5 The Business Case for Modularization.

2.2.3 Increased Productivity

As already noted in cost and schedule benefit explanations, the main driver for modularization is the increased productivity that comes with building off the site. This increase is driven by the combination of lower labor rates along with greater productivity.

The labor rates are a function of the country or area of the world. Many of the module fabrication yards are located in the Far East, SE Asia, and even the Middle East, where labor is less expensive than in the US or European countries. Of course, this benefit shrinks where the project site is also located in an area of less expensive labor. So, we have to take a look at the labor productivity factor as part of the comparison, as explained at the beginning of this chapter.

Productivity is a function of the skill of the craft workers. It is also a function of the setting where they must work, the equipment provided, and any restrictions placed on them in terms of working in a specific location. All these factors impact their total time on tools. The workers must deal with weather, working at heights, and even working in "hot" (or energized) areas at the site. All of these require additional paperwork, additional clothing and safety gear, fire watches, and other extra supervision. All this limits the time that the craft actually gets to apply to the actual "building" of the plant.

The module fabrication yard has been specifically set up to optimize all the processes required to fabricate, assemble, and ship the module, from the orderly receiving of bulk materials, through the various steel, pipe, and vessel fabrication lines of production, over to the assembly areas where these individual pieces are grouped into sub-assemblies or portions of decks, which are further combined into pancakes, and finally stacked into the multi-level modules.

As noted in the explanation above, a lot of this work is performed at grade, in areas specially set up in terms of utilities and other support needs for these different fabrication efforts. They are free of any restrictions in terms of conflicting activities and often performed indoors, so free of delays from the weather. All of this allows the craft to spend more time actually doing the work.

2.2.4 Improved Predictability (Surety/Reliability) or Less Variability

One of the key benefits of modularization is predictability (or surety and reliability) on schedule, cost, and quality. This benefit has been under-highlighted in the past. However, several leading global companies in modularization stated that project performance variability was less when the extent of modularization was high compared to the conventional stick-built project.

The module fab yard is a module-producing facility. Like any other manufacturing and assembly plant, its profitability is based on getting engineering and materials in and fabricating, assembling, and shipping completed modules out either on (or ahead of) schedule and on (or under) the budget. The profit margin is tight, and anything that delays this effort becomes very costly for the fabricator, as it not only delays the current jobs but it potentially impacts the contractual start of the subsequent jobs. So, the fabricator will try to meet the schedule as it is in the fabricator's best interest to stay on schedule. Furthermore, they will do so as long as the deliverables being provided by the owner/client or the EPC contractor are on time.

Therefore, any delays on the part of the owner/client or EPC contractor become very expensive for the project. When working a module job, it is critical in terms of both cost and schedule to make sure that the module fabricator has what they contractually need to assemble, build, and deliver. So, in the end, a module job will be on schedule if the outside deliverables to the module fabricator are on schedule. If they are not provided on time, then the results can be much worse than the equivalent delays to a stick-built project.

2.2.5 Increased Safety and Quality Performance

As noted above, with dedicated manufacturing lines as well as assembly areas, conflicting activities and work not related to the main production at hand are eliminated. Also eliminated is working at heights. With the proper lifting and handling equipment, many of the potential strains and over-exertion issues associated with having to "horse" a piece of material or equipment into place are removed. With groups of craft regularly working together, a team mentality develops where each knows what others can and will do. Fewer unexpected actions are encountered. In the fabrication yards, the safety equipment is standardized and regularly inspected. More importantly, it is associated with the work at hand.

Furthermore, the modular approach will lead to improved quality for the same reasons above. Better safety of the workers, better quality control, and a better environment with no or less dust and debris (many fab yards are paved or work on stabilized soils) all contribute to a better quality of modules and the project.

2.2.6 Increased Sustainability Performance (Green Benefits)

One of the best ways to reduce construction waste is modularization. This benefit has been under-estimated by the industry. However, increased interest in the environment, a Green and Lean philosophy has turned attention to the advantage of modularization. The modular approach not only reduces pre- and post-construction waste but also reduces the actual materials used. Furthermore, it also reduces rework, overall energy consumption, and greenhouse gases, which can all contribute to cost savings.

2.2.7 Site and Site Construction Team Benefits

In addition to the benefits of structuring equipment and piping into "modules" compared to the stick-built arrangement, there are significant benefits to the site and construction team, including:

- longer duration for foundations and underground;

- higher efficiency at the site;

- completion of pre-commissioning efforts off the site;

- less impact on the site.

2.2.7.1 Longer Duration for Foundations and Underground

Because it takes longer in terms of the overall schedule to complete the module and get it ready to ship to the construction site, the module arrives later in terms of the overall project schedule when compared to similar activities for a stick-built schedule. This allows more time to those workers at the site who must complete all the necessary underground work on foundations, underground piping, utilities, drainage, water treatment facilities, and any major earthwork.

However, note that incomplete underground efforts, if not completed in time, are detrimental to the schedule and cost of projects, especially where work has to be "covered up" and then re-opened to provide delivery access to parts of a plant. Furthermore, all the other underground activities for module tie-in must be out of the ground prior to the module's arrival. The modules, when they arrive, are essentially complete, requiring final hook-up and any additional testing of the connections. This is performed by a much smaller crew than what would be required to assemble the module scope as a stick-built endeavor.

2.2.7.2 Higher Efficiency at the Site

The benefits of a reduced workforce at the construction site go beyond the obvious replacement of higher costs and a less efficient site workforce with a less costly and potentially more efficient workforce. The smaller workforce can move to and from the site more efficiently, work in a less crowded environment, and concentrate on the specific specialties of hook-up and commissioning.

2.2.7.3 Completion of Pre-Commissioning Efforts off the Site

The aim of modularization is to complete (as in 100%) the module scope. This typically includes much of the pre-commissioning and checkout work on the equipment, electrical, instruments and control (E&I&C), as well as on all the piping. If this pre-commissioning (and potentially some commissioning) work are adequately planned and include the appropriate quality control and checks, as well as approval and sign-off by those who have the authority to approve that level of cleanliness and completeness, then there is no reason to re-work any of this effort on site. The result is a much shorter and faster commissioning and start-up effort.

2.2.7.4 Less Impact on the Site

Exporting site-based work to the fab shop/yards also helps minimize the impact on the construction site. Compared to the traditional stick-built approach that generates extensive dust, noise, and air and water pollution at the site, the modular approach greatly reduces those impacts (Choi, Chen, and Kim, 2019).

2.3 Challenges

Another word for barriers (disadvantages). Obviously, with all the good benefits that come with a modular approach, there will be some challenges. With the incorporation of modules into a project's planning and execution, there are numerous barriers that have been hinted at and danced around but not specifically stated and explained. The challenges of modularization that we address here are:

1. Critical path (reduction in flexibility)

2. Upfront cost (and pre-commitment)

3. Barriers to engineering

   a. Design must support the fabricator's planning and build methods

   b. Earlier engineering efforts

   c. Different dynamics in the engineering organization.

4. Acceleration of procurement

5. Owner and contractor capabilities

6. Module fabricator skills and capabilities

7. Extensive coordination

8. Logistics (module transportation)

9. Others

Overcoming these challenges is critical in setting up the project for success. Below, we explain these challenges. However, note that not all these challenges will necessarily apply to all the modular projects. In some cases, these barriers will vary in their extent and impact with the type of

contract, timing of decisions, cash flow, project approval, or responsibility matrix, to name a few.

2.3.1 Critical Path (Reduction in Flexibility)

Because of the need to design the module, purchase bulks and equipment, ship to an intermediate fabrication site, prefab materials, assemble, stack, load out and ship to site, the module typically lands on the project critical path. Further, because the delivery of the module is already later than the similar situation in a stick-built project, any further delays in the shipping and delivery compress the remaining schedule activities required for mechanical complete, commissioning, and ultimately start-up.

So, not only is it good business to make sure that the fabrication yard is supplied with everything it needs to deliver on schedule (to avoid extras and change orders from the fab yard), it is also necessary to avoid additional scope creep coming into the project site late in the project's life due to carryover work. With the shortened remaining time, the project cannot afford to add more scope at these late and critical phases. Thus, it is critical to freeze the scope and design early, which makes it unattractive to some owners/clients who like to continue to "tweak" the process or project details.

2.3.2 Upfront Cost (and Pre-Commitment)

One often hears that a module job will cost more than a stick-built job. We tend to agree BUT with one huge caveat—assuming all other factors between the two execution strategies are the same. But, as previously mentioned in Chapter 1, the issue is that not all of the factors that make a successful job are available on site. Many more of these are available at the module fabrication yard, so the comparison is never accurate.

A better way of putting it is to restate the challenge, saying that a module job will require more upfront money to be successful. While it sounds counterintuitive, sit back and make a mental comparison of the stick-built and module execution plans. The stick-built plan designs, purchases, and ships all the pieces to the final project site to be set up and assembled there. The module execution plan requires that everything modular gets sent to an intermediate project site (called a module fabrication facility) to be set up and assembled there. Once completed, it now must be shipped to the project site, where it is set in place and hooked up. However, this intermediate stop to assemble at the module fabrication facility requires additional time, the shipping of the completed module to site requires additional time, and the onsite connection and final hook-up after setting requires additional time. Even with the module being shipped later in the schedule than the individual pieces for the stick-built option, there are still enough slices of additional time needed to put the module on the project critical path.

To get the module off the critical path, some work must be started earlier to get the module chain of events going sooner. In many cases, this critical path runs through the structural steel design of the module. This structural steel design is based on an analysis of the equipment and loading on the members. As such, this equipment must be designed and a vendor selected in order to purchase enough engineering details to determine where the loads will be placed and, in terms of rotating equipment, where the appropriate bracing is needed to avoid excessive vibration and harmonics. The vendor is not going to provide such a design without contractual assurances that, should the project go forward, they would be awarded the job.

If the client's project approval procedures do not allow any money for pre-commitment on these critical equipment packages, then the structural steel cannot be designed and purchased. While this may be an oversimplification of the issue in some cases, it can be challenging to resolve for large and complex modules unless the client is willing to provide funds for early engineering of critical equipment.

2.3.3 Barriers to Engineering

The barriers in terms of engineering are numerous. Here, we address only the key barriers in engineering: (1) the design must support the fabricator's planning and build methods; (2) earlier engineering efforts; and (3) different dynamics in the engineering organization.

2.3.3.1 The Design Must Support the Fabricator's Planning and Build Methods

The design must support the fabricator's planning and build methods. The module fabricator has a very

well-defined production and assembly program that they follow. This program must be supplied with inputs on a definite schedule, starting with engineering design, preliminary material takeoffs, approved construction drawings and 3D model, equipment, etc. Delays in any of these activities run the risk of producing a much bigger delay in the module completion and delivery if, as a result, one or more of the sub-assembly lines at the fabrication yard cannot deliver.

This becomes even more critical as the modules become larger and more complex. Large rotating equipment and tall vessels provide a structural steel design challenge in terms of rotating dynamics and moments that must be resisted. All of this takes time to design and depends on the inputs from the equipment vendors. Late information results in late structural steel design, which results in late procurement and assembly, which ultimately results in an overall delay in module fabrication. It becomes crucial that as the module increases in size and complexity, the timing of developing all the supporting engineering needs to extend, requiring an earlier commitment from the successful vendors and manufacturers.

Often, satisfying these needs directly opposes the owner's/client's contracting strategy and approval processes. When funding required for early engineering efforts is delayed due to the contractual stipulations that provide no funding until project approval or a Financial Investment Decision (FID), then project management is saddled with the tough decision on how to proceed. If no relief is provided in terms of pre-commitment of funds for certain critical design efforts, the EPC contractor is left with the only alternative of leaving this equipment out of the module. This is a decision that can result in a module plan that is not optimal in terms of cost due to the funding schedule restraints. This affects both the owner/client as well as the EPC contractor.

2.3.3.2 Earlier Engineering Efforts

Not enough and not soon enough: The module job requires *earlier engineering efforts* on most everything that defines and goes into a module. During the typical FEED (or FEL3) phase on a stick-built project, the level of detail required for structural steel is for the supporting piping on pipe racks. This is often not critical and does not get a detailed design until well into EPC.

For the module job, especially for the large integrated modules, there is a need to understand the equipment well enough to be able to design the primary steel and, in some cases, secondary steel if vibration reduction is necessary. To do this, engineering must have the process design complete enough to actually go out for bids on the equipment, receive the bids, evaluate them, and conditionally award to the successful bidder, in order to get that bidder to provide the engineering effort of developing enough detailed design to provide the needed forces of the static equipment and the dynamics of the rotating machinery as input to the structural designer.

2.3.3.3 Different Dynamics in the Engineering Organization

What this creates is a different dynamic in the engineering organization. For the stick-built project, the piping and plot plan layout groups typically drive all the other disciplines since they are the recipient of all this information in order to develop the plot plan and layout and later route piping. For the modular job, it is the structural group that sets the schedule of what data are needed when from the following sectors:

- the equipment engineers—who must get it from vendors;

- procurement—who must work with equipment engineers of the bidding and selection process;

- piping layout—who must work with the process engineers in terms of developing exactly how the equipment can be laid out in the three-dimensional module;

- piping engineering—who must develop pipe stresses that will need to be resisted by the module steel;

- and so on.

As a result, when the module schedules are laid out within the project schedule, the module fabricators identify when they need to have the bulk and shape steel delivered (which will be the basis for the module build), procurement determines the timing of purchase and the structural engineers set the due dates for all the inputs required into their steel design. The project schedule must ensure that everything needed for the steel design is being started in time to deliver. As a result, the "extra" time the engineering team had during a stick-built job to do the following is not available for the module job:

- work out equipment layout and supports (because they were sitting on massive concrete foundations);

- develop stress calculations (because they were supported by simple pipe racks);

- identify location and routing of piping (now in a module instead of on a pipe rack or finger rack);

- identify the need and location of the switchgear/ motor control center (MCC) or remote terminal unit (RTU) building (which now may reside in a module);

- identify many other smaller issues that must be resolved in time to develop the module structural steel design.

This sets up the structural engineering discipline as the driving discipline to meet the schedule. This can be an issue with project teams that are not familiar with this hierarchy, with the result that the structural team does not get the inputs they need in time, and the entire module schedule suffers.

2.3.4 Acceleration of Procurement

Also involved in the acceleration of data required to get a proper design on the module structural steel is procurement. As previously mentioned, this group must not only accelerate their efforts in terms of three bids and a buy (or whatever bidding process they use) in order to identify the final vendors. As part of this process, they must include contractual verbiage that provides the flexibility to make an early purchase of a small percentage of the engineering required to perform the layouts, work the dynamics, develop the secondary support steel, etc., that is required to fully design this module structural steel.

This is not their standard operating procedure, and, if left to their own stick-built timing, they will not have vendors on board in time. Since much of this effort occurs prior to FID, these contracts must be set up differently to include limits in terms of total cost outlays, should the project not get funded.

The other obvious barrier to early procurement is the owner/client, who, after being shown the need to pre-invest in some of this early bidding as well as purchase some of the early equipment engineering, decides they cannot provide such pre-investment funds. If such pre-investment money is not available to perform these pre-FID activities, then the EPC contactor is left with the option of either leaving long lead equipment out of the module, overdesigning to accommodate, or reducing the size of the modules to gain schedule advantage (assuming these smaller modules can be built faster than the larger modules). Again, the result is a less than optimal module solution.

2.3.5 Owner and Contractor Capabilities

This is our favorite challenge. Because an owner/client and EPC contractor do not understand the wide-ranging impacts across all project groups by the decision to modularize, they sometimes naively assume that there is no need for a module coordinator or module team, considering it an unnecessary expense. On some projects, this belief is so strong that there is nothing you can do but try to get as far away from the project as you can. However, most project management has enough experience (including associations with one or more previous module jobs) that they will consider including the module team.

However, in either case, it is vital that the project team has members in management with past modular experience, and this experience should be from start to finish on that particular module project. There is too much interaction throughout the project, from the initial assessment of module options to the final on-site set-up, to assume one can get a complete understanding by working only on a part of it.

If the owner/client does not have a module subject matter expert, they should hire or "rent" one. The same thing goes for the EPC contractor. Furthermore, it is crucial that both the owner/client AND the EPC contractor have this experience and leadership on their respective teams. It does not help the overall project if the EPC contractor is the only one with modular experience. They will end up spending most of the time they need for coordinating, tied up trying to explain everything their company is trying to do differently or needs to be done differently. And often, as a result, this effort of "dragging the owner/client kicking and screaming to the module learning table"

will only end up obtaining some half-hearted concessions grudgingly made by the owner/client and with no strong support behind them.

The following works much better (and we have seen exactly such a scenario unfold). The owner/client representative, typically the project manager or project director, calls all their company team as well as the EPC team members together to discuss the project, goals, and challenges. In this presentation, after explaining all the difficulties with building the project in the middle of their existing operating plant, including the disruption to production, delivery of products, and access, they tell the composite team sitting in front of them that while they are no experts on modularization, they realize it is the only answer to keep plant disruption to a minimum. A such, their company has contracted the module-savvy EPC team sitting with them. And, further to this, they expect all team members to accept this option and do everything in their power to make it successful. They conclude by saying that this will need to be a complete team effort, so anyone who is not fully engaged and supportive of this project execution should excuse himself/herself from the team and leave through the door they came in.

There is just too much at stake and too many small changes in priorities required to have an undercurrent of opposition to the execution plan.

2.3.6 Module Fabricator Skills and Capabilities

Module fabricator skills must match the project needs. Module fabricators come in all shapes and sizes (and abilities). They range from a truly "mom and pop" operation set up on a vacant lot that essentially bolts together pre-fabbed structural steel and then sets pre-fabbed pipe in the resulting structure, to the fully integrated multinational operations that not only own their own pipe and steel mills, where pipes and tubulars are rolled, and plate girders are fabricated, but also have complex assembly lines to support the constant supply of prefabricated steel and piping from these mills. They turn them into finished pipe spools and structural members that are provided simultaneously to numerous internal assembly site fabrication areas so that all can be outfitted with piping and equipment to be eventually joined, via stacking of these decks (or pancakes), to be finally skidded or rolled off massive module offloading facilities (MOF) and on to one or more of their own ocean-going vessels for shipment to the ultimate project site.

Bigger is not better when looking for a suitable fabrication facility. All that expertise and capability mentioned above in the fully integrated module production facility is costly and only warranted in special cases. The module job must fit the skills of the fabrication facility. If structural welders are scarce, then the suggested welded and painted module should be redesigned as a galvanized and bolted structure. If the best that the fabrication yard can do in terms of a MOF is a reinforced earthen ramp to the water, then the module scope should be in line with that capability for shipping.

On the other side of this issue, besides being overqualified, large fabrication facilities need large complex module jobs in order to support all the other parts of their organization—such as their procurement department, vessel design and fabrication groups, pipe mill, and logistics groups. Besides the need for an extremely large open area for pipe rack fabrication, the simplicity of the assembly does not utilize these additional departments and, as such, makes the simpler modules of no interest to these fabricators. In fact, if they accept the simpler modules as part of their scope, they may immediately subcontract the pipe racks out to a lower-cost fabricator with a yard suitable for the large acreage requirements of the pipe rack assembly.

In addition, the size of the fabricator is critical. The proposed module scope needs to be in a range that is not above the fabricator's maximum capacity but large enough not to be lost in the yard or ignored. It needs to be a large enough size to require a significant percentage or even a slight majority of the fabricator's time, labor, and materials, so that they must pay attention to the project details.

Suitable fabrication facilities are reviewed and selected via a fabrication survey, a multi-page questionnaire requesting information from capabilities to the backlog to safety statistics. The responses are reviewed and, from this assessment, several are selected to receive the request for proposal (RFP).

2.3.7 Extensive Coordination

A module job requires coordination at all levels with all groups involved in the project. This can be a problem where the project is not set up to handle this coordination.

What does this coordination look like? It is essentially a small group of experts who have banded together based on their knowledge of what it takes to execute a modular job. The purpose of this group is to interact with all the other groups within that project to make sure that decisions made take into consideration the modular aspect and educate the other groups on what is expected and when it is expected. The details of this team can vary, but the location of the module team lead in the organization is critical. See

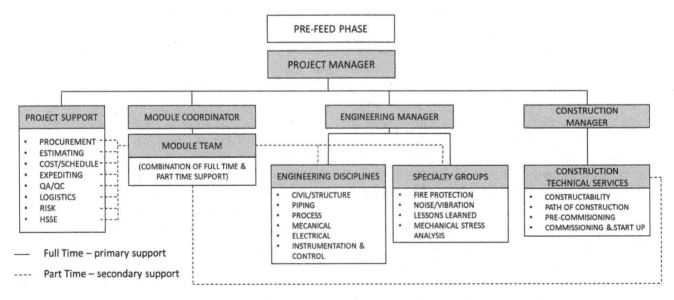

Figure 2.3 A typical pre-FEED phase organization chart for modularization.

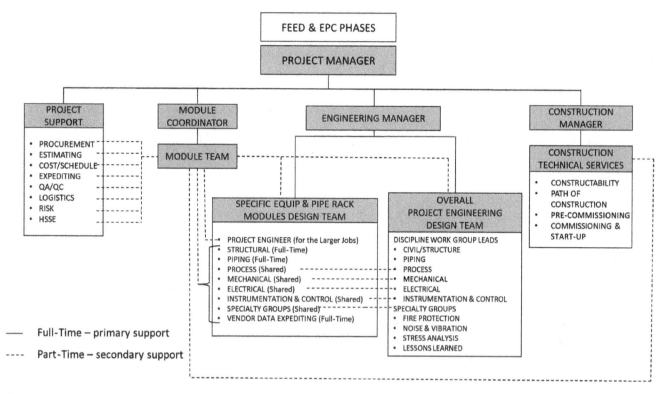

Figure 2.4 A typical FEED and EPC phases organization chart for modularization.

Figure 2.3. for typical pre-FEED phase modularization team organization and Figure 2.4. for FEED and EPC phases.

The module coordination team is composed of a module coordinator with representatives from each of the major groups within the project. It is more like a matrix arrangement, as each appointed member for a group must also support the module team activities.

The module coordinator is set up in the organization to report to the project manager at the same level as the engineering and construction managers. This allows the module coordinator to keep the module activities on an equal basis with the engineering and construction activities and avoids any conflicts in priorities that might occur if the module coordinator was reporting to either the engineering manager or construction manager. The module coordinator has a dotted line tie to both the project support groups as well as the engineering design groups and is, in effect, another input to these groups in terms of module decisions.

2.3.8 Logistics (Module Transportation)

This is such a big issue that it typically receives the early analysis that it should. As one of the very first actions in a

module design development, the transportation/logistics/route survey will identify the maximum size module that can be moved from one of several fabrication facilities to the project site. The larger the module, the more difficult it is to find a project site that will have an access route of sufficient clearance to accommodate it. Also, the larger the module, the more difficult it is to find adequate lift cranes (if self-propelled modular transporters [SPMTs] are not being used to set the module) as there is a worldwide shortage of mega lift cranes, which can be costly.

This only becomes a barrier or challenge if the module size changes, the route clearances change, or there is a bust on the route survey analysis. More commonly, projects decide to skip this survey because they do not think they will modularize; they had one performed recently; they do not think it is necessary; or they just do not think it is worth it. As a result, the "recent" survey (of 10 years ago) that is the basis of the project design was not new enough to catch the more recently installed electrical transmission lines. So, the route everyone thought was clear now has owner access restrictions or actual physical structures limiting dimensional access. Also, what few people understand is that a local area Heavy Haul Contactor knows much more about the project area than the project team and usually is able to provide transportation alternatives that were not even considered options by the project team.

2.3.9 Others

This is a big catch-all for everything else, most of which has been hinted at above:

- **Management adjustments:** the ability to learn what changes are needed and adjust as required.

- **Planning:** the ability to plan around the critical path, which is typically the module production and delivery.

- **Hook-up tolerances:** the need to fabricate and install to dimensional standards that will permit a single weld hook-up (SWHU) when the two modules are mated on site.

- **Special Quality Assurance/Quality Control (QA/QC) procedures:** at the fabrication yard involving the proper paperwork signed by the responsible parties signifying cleaning, tests, etc., have been adequately performed and do not need to be re-done at site.

2.4 Summary

This chapter explained the key characteristics of fabrication yards to better understand why they can offer advantages. Then we discussed the major advantages and challenges (or disadvantages) that should be included in the overall analysis of the module potential. Note that some advantages and challenges are adopted from (Choi, 2014).

References

Choi, J.O. (2014) Links between Modularization Critical Success Factors and Project Performance. Ph.D. dissertation. The University of Texas at Austin. Available at: https://repositories.lib.utexas.edu/handle/2152/25030.

Choi, J.O., Chen, X.B., and Kim, T.W. (2019) Opportunities and Challenges of Modular Methods in Dense Urban Environment. *International Journal of Construction Management*, 19(2), pp. 93–105. doi:10.1080/15623599.2017.1382093.

chapter 3 Industry Status on Modularization

This chapter describes the industry status of modularization through a compilation of 25 actual case projects. As a quick reminder, Chapter 1 provided a basis for what modularization is (and what it is not) and where it might be applicable. In Chapter 2, the advantages and disadvantages of modularization were explained. This set the stage for Chapter 3 to describe the industry status on modularization, and this chapter validates the advantages and impediments which were initially identified in Chapter 2 via this set of actual project case histories.

Incorporated into this chapter are some of the data and findings from Dr. Choi's PhD dissertation (Choi, 2014) but reanalyzed, restructured, and summarized with a focus on the industry status of modularization rather than the original premise of Dr. Choi's dissertation, which focused on identifying the correlations between modularization critical success factors and project performance.

In particular, this chapter reports on the findings from these 25 project case histories in terms of identifying the advantages of modularization, quantifying the schedule and cost savings compared to the alternative stick-built project philosophy, recognizing the project difficulties and impediments when approaching a modular project application, identifying some of the business case drivers for modularization, and presenting a quick overview of the metrics of the 25 projects studied—the types of modules, the variability in number, size, and weight of the modules, and some characteristics of both the project job site and the module fabrication/assembly shops.

We hope this chapter will be informative as well as thought-provoking. Remember, this is actual data from completed module projects, with responders providing the data from the point of looking back at the entire project. So, while it is accurate in terms of a post-mortem review, it is still conceivably laced with some biases and preconceptions of the responder. As such, along with this presentation of the hard data and raw results, we reflect on and offer other plausible interpretations or alternative thoughts where these raw statistical results might be misleading.

At the end of the day, the data is only a snapshot—and should be taken as such.

3.1 Modular Projects Case Study

3.1.1 Case Study Methodologies

Descriptive research methodologies were applied to understand the characteristics and status of modularization. The collected data is graphically displayed via frequency distributions and bar graphs/histograms, along with central tendency (mean, median, and mode) and variation (range and standard deviation) per item (Choi, 2014).

3.1.2 Sample Characteristics

In short, 25 sample projects were collected in 2013 from modular subject matter experts who had played a major role in the sample projects being surveyed. The average percentage modularization of the sample projects was

52%. The types of modular projects and locations of the project sites and fabrication/assembly shops are summarized below.

3.1.2.1 Types of Sample Modular Projects

Figure 3.1 illustrates the variety of the collected industrial modular project information in terms of their diverse nature, with representation from mining, natural gas processing, chemical plants, consumer products manufacturing, oil

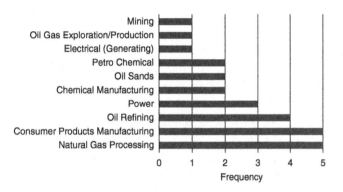

Figure 3.1 Types of collected industrial projects. Source: Choi (2014). Reproduced with permission of Jin Ouk Choi.

refining, and power. The point of this illustration is to demonstrate that the results crossed many project types and, as such, should be considered to be more or less universally applicable to all types of modularized projects.

3.1.2.2 Project Job Site and Fabrication/Assembly Shop Location(s)

The projects surveyed were truly worldwide, as shown in Figure 3.2, where the 25 projects were plotted in terms of their job site, module fabrication shop, and assembly shop (where different).

Project job site location and module fabrication and assembly shop locations for the collected industrial modular projects were assessed and grouped by major country or continent (see Figures 3.3 and 3.4).

As expected, the majority of the module fabrication assembly is performed overseas—only 8% of the sample projects' job sites were located in Asia, but approximately 29% of these projects were contracted to module fabrication and assembly shops located in Asia. Considering the timing of these surveys (2013) on projects completed

Figure 3.2 Location of job site, module fab shop, and assembly shop. Source: Choi (2014). Reproduced with permission of Jin Ouk Choi.

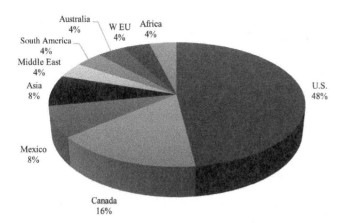

Figure 3.3 Project job site locations. Source: Choi (2014). Reproduced with permission of Jin Ouk Choi.

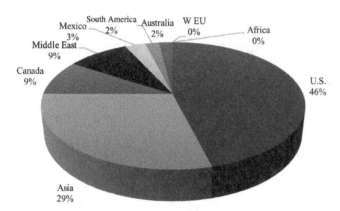

Figure 3.4 Module fabrication and assembly shop locations. Source: Choi (2014). Reproduced with permission of Jin Ouk Choi.

by that time, this large percentage of fabrication effort in Asia was probably due to the potentially large cost savings that could be realized in terms of the all-in wage rates (AIWR) differences between the Far East craft workers and the US craft workers, where most of these projects were being built. Even the slightly lower difference in craft labor productivity between the two areas was not enough to offset the tremendous cost savings.

As a cautionary note, while there were potentially large differences in the AIWR between the Far East and the US, it should be understood that even within a given area, like the Far East, both the AIWR and craft productivity varied widely between different areas and in some cases even between shops within the same area. So, with over 200 module fabrication shops worldwide, finding the best fit between the project and the fab yard is not just a simple selection of an area. As such, it can be a daunting effort

that must be approached properly and examined adequately. Later chapters go into this effort in more detail.

3.2 Results (Industry Status on Modularization)

3.2.1 Advantages of Modularization

One aim of this survey was to identify the main reasons why a project would decide to modularize. The survey respondents were requested to identify (from a list of potential advantages) all that applied to their specific project. Since the survey respondents were not limited to selecting the top 2 or 3, they provided a listing of all that applied, with over 100 responses to be tallied. Figure 3.5 illustrates the frequency of selection of each of the suggested benefits provided in the listing.

As expected, schedule and cost savings were the predominate reasons for modularization. This was followed closely by increased safety and increased productivity. The rest of the selected benefits seemed to be somewhat evenly divided in terms of percentage (1% or 4% or 6%) of times these projects identified this as a benefit.

So, what might this mean? It could mean that only a few project respondents felt that there were other project benefits from modularization other than the top four

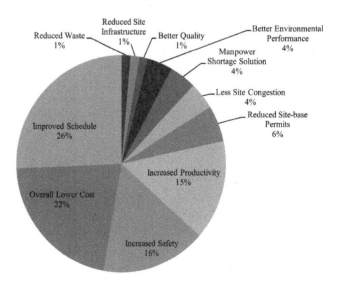

Figure 3.5 Advantages from modularization application in industrial modular projects. Source: Choi (2014). Reproduced with permission of Jin Ouk Choi.

(schedule; cost; safety; productivity) listed above. It could also mean that the rest of these surveyed projects did not receive an actual benefit from the more subtle benefits shown in the categories that received 6% or less. But, the fact that they are so evenly divided up probably means that only a few of the projects actually took the time to examine the entire modularization process and evaluate all the benefits that their project actually realized. The projects that did so were probably the larger projects in terms of size or cost where these more subtle benefits became glaringly obvious due to their physical size or impact of the project on the job site area.

From the study work on the benefits of modularization identified in the CII RT-283 research (O'Connor, O'Brien, and Choi, 2013) as well as the two authors' experiences with modular projects, we know that some projects select modularization as a project philosophy even where the cost or schedule savings, at face value, maybe minimal to none when compared to the cost or schedule utilizing a stick-built execution philosophy. But, in many of these cases, such projects select modularization for one of the many other benefits. Examples include:

- Additions to an operating plant where the construction disruption would be extremely detrimental to maintaining the current product production and sale.

- Getting a jump on the schedule with the ability to design, fabricate, and assemble parts of the plant prior to the approval of the on-site construction permit.

- The opportunity to progress a project faster than by using on-site labor where there is an inability to obtain or financially support sufficient craft labor on site.

- The ability to coexist harmoniously with local communities by reducing the increased congestion, noise, or immediate impact from the large influx of temporary workers required to completely stick-built the project.

All of these can become even more critical than the overall cost and schedule benefits, and, as a result, if these are not adequately addressed, they could jeopardize the entire project.

So, the modular philosophy provides all of the above advantages—it is just that some projects do not recognize many of them because the benefit may not be critical (or even noticeable) to the success of the project. But for those projects that must meet one or more of these "lesser" benefits, the modularization philosophy offers the ability to have greater control over such issues. As such, many companies select modularization for reasons other than the cost and schedule savings compared to stick-built. They recognize that these softer risks (obtaining permits, congestion, workforce availability, environmental impact, and reduced infrastructure) are indeed more controllable and reliable for the modular project than the stick-built alternative.Further analysis of the survey data was conducted based on what the project planners identified as the primary project driver between cost/schedule/ balanced drivers. Figure 3.6 illustrates the various project driver splits in terms of the identified modularization advantages when evaluated in terms of these three primary project drivers.

For the cost-driven projects, obviously, cost was of major importance. As expected and illustrated in Figure 3.6,

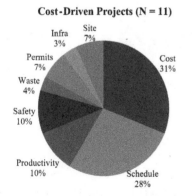
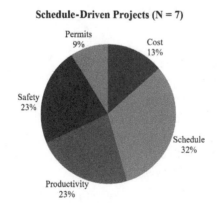
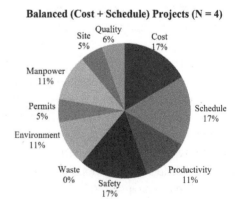

Figure 3.6 Advantages of modularization by primary project driver. Source: Choi (2014). Reproduced with permission of Jin Ouk Choi.

the most common advantages from modularization application on the cost-driven projects (N = 11) were overall lower cost and an emphasis on schedule (nearly 60%). But, to get this lower cost, both schedule and productivity were not emphasized as much as the schedule-driven project. Also, surprisingly, safety did not seem to be as predominant in terms of identified importance. What does this mean? The lower-cost job was apparently willing to accept a lower productivity from a cheaper labor force and take the anticipated schedule hit. Of course, we do not conclude that safety actually took a back seat in terms of these jobs. But we do see that the other factors played a more predominant role in the minds of the project planners.

For the schedule-driven project, obviously, the cost had to take a back seat. The most common advantages from the schedule-driven projects were schedule, increased productivity, and safety (N = 7). This makes sense because with an increased schedule emphasis comes a higher grade of craft labor to obtain a higher productivity (to get things done faster). Safety is much more important because of the positive benefits of a happy and safe working workforce. But also conversely, because of the potential negative impact that a poor safety performance could have in terms of slowdowns, stand-downs, and other potential productivity impacts.

For the balanced project, we see that all of the proposed benefits must be juggled to maintain the fine line between a cost-efficient execution and an aggressive schedule. So, craft labor costs need to be watched as well as their productivity with efforts to maintain an efficient but not costly workforce. Same thing with quality, waste, and environment. All of these must be balanced to avoid excesses in terms of one or the other, which would then impact either cost or schedule, or both. Of course, this is the hardest of the three project drivers to maintain; but because it is so difficult, it is the best suited for a modular project execution philosophy. It requires the recognition and utilization of most of the benefits that the modular execution philosophy can provide.

3.2.2 Cost and Schedule Savings Compared to Stick-Built

The case study data were also reanalyzed in terms of the approximate percentage schedule and cost savings compared to stick-built. Figure 3.7 illustrates the percentage schedule and cost savings compared to stick-built. The average percentage schedule savings was 12.5%, and the average percentage cost savings was 19.2%. The median percentage schedule savings was 9%, and the median percentage cost savings was 20%.

These findings aligned with the other literature well, where cost savings were reported about 15% (Rogan, Lawson, and Bates-Brkljac, 2000; Post, 2010), including

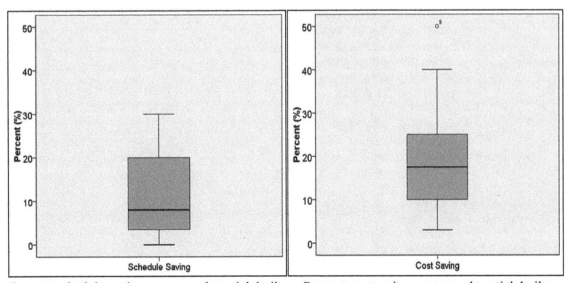

Percent schedule savings compared to stick built   Percent cost savings compared to stick built

Figure 3.7 Percentage schedule and cost savings compared to stick-built. Source: Choi (2014). Reproduced with permission of Jin Ouk Choi.

18% savings from a modular gas-oil hydro-treater project (Jameson, 2007) and savings of 15% from a solid fuel-fired facility modularization (Gotlieb, Stringfellow, and Rice, 2001). Furthermore, the recent McKinsey & Company report (Bertram *et al.*, 2019) claimed that modular construction could cut costs by up to 20% and duration by as much as 50%.

3.2.3 Recognized Project Difficulties and Impediments to Modularization Application

To understand common project difficulties in industrial modular projects, the respondents were asked to report the project difficulties that they recognized as leading to adding cost or delay. Drawing from the literature review, a list of difficulties was drawn up and provided in the survey questionnaire with instructions to respondents to check all those that applied. Those difficulties were:

- contract terms
- weather (extreme)
- logistics challenges (transportation of modules)
- environmental impact
- organizational change
- scope change
- labor issues
- regulating impact
- external stakeholders
- material shortage
- major quality problems
- change in demand for product
- change in project profitability
- change in financing environment
- safety incident
- equipment delivery
- team turnover
- and others.

Figure 3.8 illustrates the frequency of project difficulties.

The four most common difficulties were logistics challenges, equipment delivery, extreme weather, and scope change. This makes sense. If you include in the definition

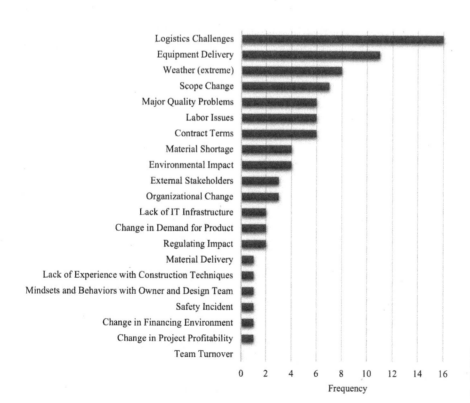

Figure 3.8 Recognized project difficulties during modular plants implementation. Source: Choi (2014). Reproduced with permission of Jin Ouk Choi.

of logistics challenges all the issues associated with module design (in terms of determining the size and ability to efficiently design the modules within the shipping confines, dictated by the site location and path to it), along with the actual issues of the logistics directly associated with the transportation, movement, and final setting-up of these modules, there are a lot of potential issues that could and do crop up.

What is interesting about the other three is they are related and, as such, can probably be grouped together. Think about it. All three are associated with the delay or impedance of the module fabrication effort. Equipment, in terms of delivery, setting, and fitting in the module. Weather in terms of fabrication slowdowns (but also could relate to the actual site weather issues). And, finally, and, most importantly, scope changes—the second most damaging impact to a module fabricator and their cost and schedule, only slightly less impactful than the actual late delivery of IFC (issued for construction) design documents or the late delivery of bulk steel and piping materials that the fabricator needs to get started.

The next three also share a camaraderie of sorts—quality problems/labor issues/contract terms. Quality issues are a result of poor materials/poor prefabrication efforts/poor assembly, all of which are related to either poor supervision or poor execution. The same can be said of labor issues—most seem to be associated with poor supervision or poor direction and sometimes can be traced back to a fabricator that is spread too thin and trying to work without sufficiently experienced first-line supervision.

The contract terms are associated with this group only because flexibility is lacking in the contract in terms of being able to address and rectify the other two issues. It may be due to an unequal sharing of risk or no contractual vehicle to adjust to unforeseen changes to the other two. In any case, with the module sitting in the fabricator's yard, the owner and EPC contractor are not in the driver's seat when it comes to any contractual coercion. The module fabricator has the upper hand in any contract dealings. As such, it is best if the project starts with a contract that is a fair and balanced approach to the project for all sides.In addition to identifying the most common difficulties of applying modularization, respondents were also asked to review a list of project impediments and identify which ones pertained to their module project. The list of impediments provided to the respondents were:

- initial cost investment

- coordination

- anti-module-oriented design

- heavy lift

- owner capability/tendency,

- contractor capability

- fabricator capability

- logistics

- shipping limits

- design freeze

- transport restrictions

- others.

Figure 3.9 illustrates the frequency of reported impediments of modularization application.

The five most common impediments were

- owner capability/tendency

- lack of design freeze

- coordination

- shipping limits

- transport restrictions.

The first two were equally ranked—one refers to the owner and the other to the EPC contractor, with both of them referring to each other's collective inability to perform. For the owner, it is typically a lack of being module "savvy" and, therefore, sometimes being more of a detriment than a benefit when working on the project. For the EPC contractor, it is the lack of a timely design freeze, again, showing their collective lack of understanding of the impacts on the module fabrication effort that late changes have and a lack of discipline in terms of a rigorous change management procedure. The third one is, interestingly enough, "Coordination." This single word combines all of the failings of the owner and EPC contractor previously noted above but is now compounded

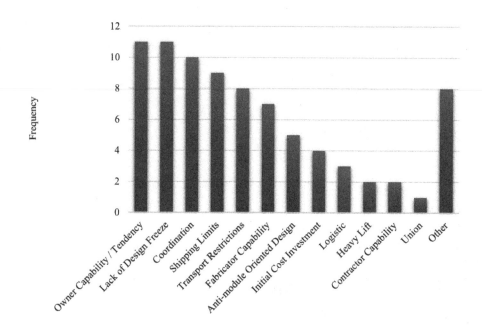

Figure 3.9 Impediments of modularization application. Source: Choi (2014). Reproduced with permission of Jin Ouk Choi.

in terms of poor communication and lack of a unified approach to the project.

The next four are an interesting mix: shipping limits/transport restrictions/fabricator capability/anti-module oriented design. Shipping limits and transport restrictions need to be identified very early. If not, they will continue to be a problem. Assuming these are resolved, a bad combination of project needs and fabricator capabilities can become a major problem if not identified in the fabricator reviews and solicitation efforts. Finally, a design team that is not in tune with how a module is fabricated and built will create problems not identified until at the fab yard. Another reason to get the module fabricator involved early is to provide a "module-friendly" design.

Other reported impediments for modularization were local labor requirements, materials management, vendor data, and timely IFC documents from engineering, quality control, an EPC tendency to build it non-modular, and government regulations. One responder highlighted that global modularization is currently constrained by the various government regulations and restrictions, units, standards, shipping limits, and data transfer regulations.

3.2.4 Business Case Drivers for Modularization

Business case drivers for modularization are defined as factors that impact the business case of a project in terms of the project objectives and explain why the modular project philosophy might be required. These factors include:

- schedule
- labor cost
- labor productivity
- labor supply
- safety
- quality
- environmental issues
- regulatory issues
- legal issues
- site access
- site attributes
- security/confidentiality
- sustainability
- predictability/reliability
- disruption (Choi, Chen, and Kim, 2019).

Figure 3.10 plots the business case drivers in terms of the most often identified reason for implementing a modular philosophy on a project to the least often identified for this sample of 25 projects.

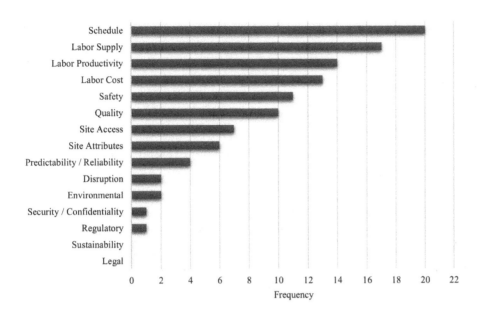

Figure 3.10 Business case drivers for modularization in the industrial projects. Source: Choi (2014). Reproduced with permission of Jin Ouk Choi.

What is interesting about this list is that the schedule benefit is at the top! This seems to fly in the face of all those who espouse the opinion that a module philosophy will result in a longer schedule. What typically is the confusion about schedule is that when looking at the engineering/design/module fabrication/shipping/setting linked tasks of a project and their duration, one can make a case for this set of tasks being slightly longer than the equivalent stick-built durations (due in part to the additional early study work as well as the duration of the completed module shipment). However, while this may be the case for this small string of activities, properly placing this segment of work within the total EPC project schedule will result in an overall reduction in project duration. This is because the modules can be shipped later in the site construction duration, and in some cases, the site duration can be reduced due to less scope required to be completed at the site. The challenge is to avoid taking the typical stick-built schedule durations and sequencing and simply dumping the above module sequence into it. If this is done, then you gain no advantage of parallel work at the fab yard and site and do not benefit from the appropriate site duration reductions for the reduced scope of work at the site. In fact, you penalize both.

Labor supply, labor productivity, and labor cost are all related and pertain to the typical problems associated with the stick-built effort at the project site. Most projects continue to have to deal with a less than adequate labor supply, poorer than anticipated labor productivity, and a higher than planned labor cost. Add all of these up together, and this spells major cost and schedule implications, which in many cases were not anticipated (project management almost always likes to believe their project will be the exception rather than the rule when it comes to labor issues). Safety and quality are well known to be "better" at the module fab yard. While such safety and quality are also possible at the construction site, they are harder to achieve. Finally, site access and site attributes are typically brought up in terms of their restrictions at the chosen project site. Reducing the number of craft and personnel making that trip to the project site improves the project's ability to meet its goals.

3.2.5 Types of Module Units

To determine the type and distribution of module sizes in the 25 projects surveyed, the survey respondents were asked to report all types of module units/sub-units that had been modularized based on the list provided:

• process equipment

• loaded pipe rack

• utility equipment

• structural modules

• dressed vessels

• buildings

- power distribution centers (PDC)

- remote instrument buildings (RIB)

- power generation equipment,

- others.

Figure 3.11 shows the results of this survey.

The five most common types of modularized units/ sub-units are, as expected: process equipment, loaded pipe racks, utility equipment, structural module, and dressed-up vessels. Of interest was the observation that while pipe racks are what have always been assumed to be the most universal and easiest type of modularization, for this case study, the equipment modularization was more prevalent. This may be due to the fact that most of the case projects were on larger and more complex projects where equipment was always planned to be modularized. However, one would have expected pipe racks to also be included in every one of the 25 case studies. But, since the mix also includes manufacturing, the typical pipe rack may not have been specifically identified as a separate module type since piping is typically incorporated into some of the other building supports as part of an integrated on-site build.

Other miscellaneous reported modules were electrical sub-stations, conveyor towers and components, and major equipment with auxiliaries, such as power generation equipment sets. Some of the respondents mentioned that their most commonly modularized pieces of plant were the power distribution centers and dressed-up

vessels. And it should be noted that since there are numerous variations in the terminology for some of the more common pre-assemblies, the exact split of the units/ sub-units by category may vary depending on the type of project.

3.2.6 Number, Size, and Weight of Modules

In order to better understand the scope and variation of the projects represented by the case studies, we took a look at some of the physical metrics of the projects in terms of the total number of modules and total tonnage, as well as identifying both the largest and heaviest module as well as the smallest and lightest. The results are shown in Figure 3.12.

The top left table and bar chart in Figure 3.12 show a variation in the total number of modules in the sample projects from 1 to 500 modules with a median of 37 modules. Of these, 33% of the projects were composed of 9 or fewer modules, and 63% were less than 99. So, the smaller projects were well represented in this survey.

The top right table and bar chart in Figure 3.12 list information on the total weight of all the modules in the project and ranged from a project with a total low of 10 metric tons to a high of 181,440 metric tons, with a median of 3,895 metric tons (skewed by the few extremely large modular projects in the study). Of interest, once again, are the large number of smaller module projects that incorporated modularization into their project philosophy— 36% less than 1000 total metric tons and 64% less than

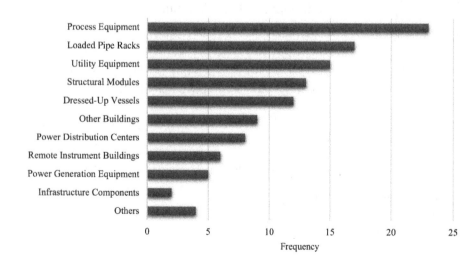

Figure 3.11 Types of modularized elements/ units of industrial modular projects. Source: Choi (2014). Reproduced with permission of Jin Ouk Choi.

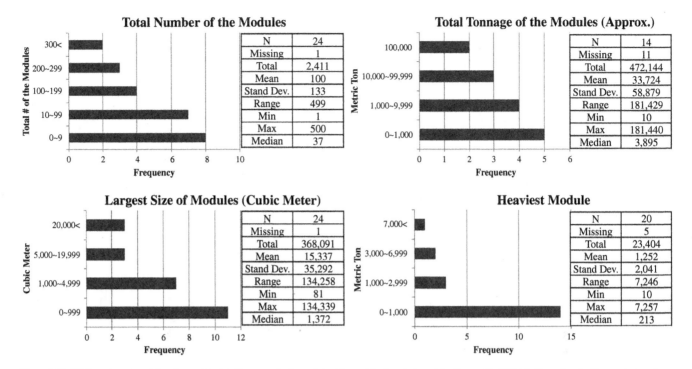

Figure 3.12 Total number of modules, the module total tonnage, the largest module, and the heaviest module. Source: Choi (2014). Reproduced with permission of Jin Ouk Choi.

10,000 total metric tons of the case studies providing details on this metric.

In terms of the largest size of module in each of the case study projects sampled (the bottom left table and bar chart in Figure 3.12), results ranged from 81 to 134,339 cubic meters. To help put this volume into perspective, the 81 cubic meter module is a typical truckable module of 3 meters wide and high by 9 meters long. The largest module is in the range of 40 meters wide by 40 meters tall by 84 meters long. So, 46% of the projects surveyed were modularizing their project using truckable sized modules.

Finally, the last metric reviewed was the heaviest module weight. Results are shown in the bottom right table and bar chart in Figure 3.12 and range from an individual module low weight of 10 metric tons to an individual module high weight of 7257 metric tons. Again, what is interesting is the lower summary range, where 70% of the projects utilized modules weighing less than 1000 metric tons. A further split of this lower range would show that the majority of these were on the lower end of this range, as indicated by the large number of smaller 99 cubic meter modules identified in the bottom left table.

So, while the numbers are impressive in terms of the maximum number of modules, maximum size, and weight of individual modules, one must remember that the majority of the jobs used much fewer and smaller modules. The module philosophy being presented throughout this book applies regardless of the size and scope. The same approach is used in determining when to start the planning, what to modularize, and how to modularize. All of these details will be discussed further in the following chapters.

3.2.7 Characteristics of Job Site and Module Fabrication/Assembly Shops

The survey also looked at the different characteristics of the job site and the module fabrication and assembly shops. In particular, the difference in labor availability, labor quality, expected labor productivity, and actual labor productivity between the job site and the modular fabrication/assembly shops. Figure 3.13 shows the results of this section of the survey.

As expected, most sample modular projects indicated that the labor supply was not adequate at the job site, whereas fabrication/assembly shops had an adequate

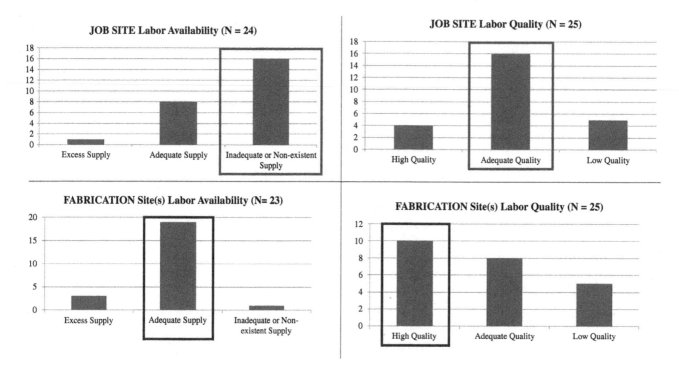

Figure 3.13 Labor availability and quality at the job sites and fabrication shops. Source: Choi (2014). Reproduced with permission of Jin Ouk Choi.

supply (the top and bottom left figures in Figure 3.13). This is to be expected since all the jobs being surveyed were module jobs and, as such, had already made the decision to modularize based on one or more key ingredients missing at the job site, most notably an adequately sized or skilled workforce. Interestingly, most of the modular projects sampled indicated that their job site's labor quality was adequate but either non-existent or not large enough to support the project construction efforts. In contrast, the fabrication/assembly shops had a sufficiently high(er) quality than adequate (as shown in the top and bottom bar graphs in Figure 3.13).

In terms of a post-mortem on these sample projects, the project study responders reported that not only did they expect "worse" or "far worse" than average labor productivity at their job site, but they were further disappointed during the course of the project. A whopping 42% of them ended up with a job site labor force that was even worse than they had pessimistically assumed, as noted by the actual job site statistics shown. In contrast, only 8% of the projects studied actually felt they had a job site workforce that was better than expected.

There was a mixed bag of results with respect to the expected and actual module fabrication site labor productivity (Figure 3.14). While 50–60% of the surveyed projects felt that the expected and actual module fab yard productivity was average or better and met or exceeded expectations, there was a sizable percentage of surveyed projects that were disappointed in the module fabrication yard productivity. Approximately 37% (or 8 surveyed projects) expected a worse than average labor productivity going into the project, and the same number ended up with actual productivity that was below expectations. We are not sure if this was a coincidence, but as we further explore how to properly set up a module project execution, we will identify some of the issues that can make a module fab yard run late with lower than anticipated productivity. Most of these issues have to do with not providing the module fab yard with either a completed design or properly detailed and accurate material take-off, late equipment, or late changes. The lack of detail in the survey makes such correlation difficult, so this metric is just left as one that needs to be identified, and the project needs to make sure that the issues that might slow a fab yard down have been adequately addressed in time.

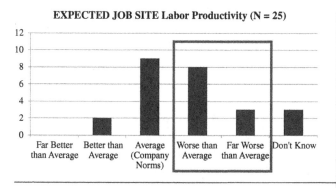

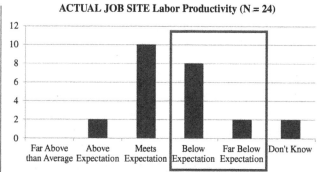

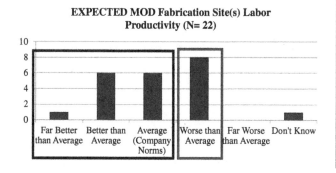

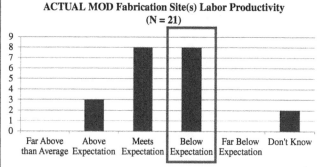

Figure 3.14 Expected/actual labor productivity at the job sites and fabrication shops. Source: Choi (2014). Reproduced with permission of Jin Ouk Choi.

3.3 Summary

It is hoped that this brief alternative analysis of these 25 project case studies in terms of their cost and schedule assumptions and expectations and their identified difficulties and impediments, along with the subsequent editorial comments regarding some of the results themselves, have provided a better understanding of these metrics. We hope this chapter has set the stage for the following chapters, which address how to properly set up and execute a modular project in more detail.

Acknowledgments

The authors wish to thank all the people involved in this research and the experts who participated in the surveys for sharing their time and expertise. In particular, the Construction Industry Institute (CII) Modularization Community for Business Advancement (MCBA) provided much of the support and leg work involved in chasing all these project case studies and gathering the raw data.

The results reflect part of the case study findings from the Ph.D. dissertation by Jin Ouk Choi (Choi, 2014) at the University of Texas at Austin. To learn more about the study methodology and findings, please refer to (Choi, 2014).

References

Bertram, N. *et al.* (2019) Modular Construction: From Projects to Products. Available at: https://www.mckinsey.com/business-functions/operations/our-insights/modular-construction-from-projects-to-products (accessed: March 16, *-2022).

Choi, J.O. (2014) Links between Modularization Critical Success Factors and Project Performance. Ph.D. dissertation. The University of Texas at Austin. Available at: https://repositories.lib.utexas.edu/handle/2152/25030.

Choi, J.O., Chen, X.B. and Kim, T.W. (2019) Opportunities and Challenges of Modular Methods in Dense Urban Environment. *International Journal of Construction Management*, 19(2), pp. 93–105. doi: 10.1080/15623599.2017.1382093.

Gotlieb, J., Stringfellow, T., and Rice, R. (2001) Power Plant Design Taking Full Advantage of Modularization. *Power Engineering*, 105, p. 31. Available at: http://www.power-eng.com/articles/print/volume-105/issue-6/features/power-plant-design-taking-full-advantage-of-modularization.html.

Jameson, P. (2007) Is Modularization Right for Your Project? *Hydrocarbon Processing*, pp. 47–53. Available at: http://content.epnet.com/ContentServer.asp?T=P& P=AN& K=27896838& EbscoContent=dGJyMNLe80Sep7A4v%2B bwOLCmr0qeqLBSsa64SLaWxWXS&ContentCustomer= dGJyMPGuslGvqbZRuePfgeyx%2BEu3q64A&D=a9h.

O'Connor, J.T., O'Brien, W.J. and Choi, J.O. (2013) *Industrial Modularization: How to Optimize; How to Maximize*. Austin, TX: The University of Texas at Austin: Construction Industry Institute.

Post, N.M. (2010) Racking Up Big Points for Prefab. *ENR*, pp. 74–77.

Rogan, A.L., Lawson, R.M. and Bates-Brkljac, N. (2000) *Value and Benefits Assessment of Modular Construction*. Silwood Park, Ascot, Berkshire: The Steel Construction Institute.

chapter 4 What Is a Module?

This chapter will introduce typical industrial module types and things to consider in their sizing. If you did not notice by now, we defined various key definitions in Chapter 1, including the term "module," but we did little more than provide a very simplified and high-level definition of the term. Now that we have spent some time with you on the advantages/disadvantages of modularization as well as the industry status on its use, we probably ought to step back a couple of steps and further define this term, as it will help you navigate through the various conversations and reading materials that discuss them.

4.1 Common Terms

The term "module," like the term "modularization," has so much connotation and baggage that comes with it that we thought it best if we spent a bit more time on it. Perhaps a whole chapter may seem like a bit of overkill, but the subject deserves more than just a few lines in Chapter 1. Since the term "module" is so generic, let's start with many of the terms used to further differentiate this general term.

First, we adopt the definitions of 'module' from Tatum, Vanegas, and Williams (1987); Construction Industry Institute (2013); O'Connor, O'Brien, and Choi (2013), which focus on industrial capital projects:

Portion of plant fully fabricated, assembled, and tested away from the final site placement, in so far as is practical.

(Construction Industry Institute, 2013; O'Connor, O'Brien, and Choi, 2013)

A module is a major section of plant resulting from a series of remote assembly operations and may include portions of many systems.

(Tatum, Vanegas, and Williams, 1987)

If you are from the building industry, when someone uses the term "module," you immediately think of small mass-produced sub/prefabricated parts/units that are basic building blocks of a manufacturing process. While the definition of "module" from the building sector theoretically also describes what we are calling a module in the industrial application, we want to make sure that this very limited definition is not carried forward because it is so limiting.

In terms of definition development, it seems that the industry has developed a set of abbreviations or three-letter acronyms (TLAs) (i.e., PARs, PAUs, VAUs, and PASs) to help further describe the various types of modules. But before we start talking about TLAs, a short apology. We sometimes can get so wrapped up in our explanations that in an effort to get it all out and on paper or in a presentation as quickly as we can, we fall back on the use of abbreviations, TLAs, or nicknames that are part of the industry vernacular we use on a daily basis. If you go to the back of this book, you will find we have dedicated an entire chapter to these abbreviations.

Why TLAs and nicknames? Because they are easy to use and every industry has them. The construction industry uses a bunch. Even worse than the construction industry is the oilfield (where one of us came from). There is literally a TLA, nickname, or common name for almost everything in the oilfield, from the well to the rig, from drilling to the day-to-day operations. A few that immediately come to mind in terms of drilling are: monkey board, kelly, kelly bushing, rat hole, mouse hole, rotary table, casing tongs, top drive, shaker, mud, mud logger, mud engineer, blowout preventer (BOP), dog house, Vee door, dogleg, fish, kick, trip out of hole (TOOH), catwalk, and stand.

Why do we bring this up? It is easy to get caught up in their use and lose an audience. So, before we lose you, we need to identify some of the more common terms we will use to describe modules throughout the book, their TLAs, generic labels, and why they exist.

4.1.1 Pre-Assembled Pipe Rack or Pre-Assembled Rack (PAR)

A pre-assembled pipe rack or pre-assembled rack (PAR) is commonly called a "pipe rack module." This type of module is the pre-assembled duplicate of the pipe racks you see in any petrochemical or process plant. It typically supports piping, power, electrical, and instrument cables in trays, becoming the linking bridge for all of these in a plant from within a process unit to other process units and eventually out to the product distribution areas.

The difference between the module version and the stick-built version is that the PAR is designed to be built off the site and, as such, must have a rack structure that is rigid and stout enough to withstand the transportation forces or accelerations that might be imposed on it while in transport to the project site from the fab yard. Also, because it has been built as a single large unit, considerations must be made with respect to its structural frame to provide structural support via the bottom girders, if the PAR is to be trailered or rolled into position, or from the top of the columns if the PAR is to be lifted and set into position via a crane.

4.1.2 Pre-Assembled Units (PAU)

A set of pre-assembled units (PAU) is commonly called an "equipment module." The relationship between the PAU and its stick-built counterpart is not as clear-cut. In fact, one of the biggest mistakes one can make when trying to modularize equipment areas is to assume that all you need to do is put a "box" around the stick-built portion of the plant, leaving the plant equipment layout as is.

This is a very inefficient module solution. The stick-built plant does not have to worry about distances between equipment pieces, and as such, the plant layout is very spacious. In the module environment, space is critical as every extra foot or meter of the footprint will require additional support steel for the PAU structural box. Also, because the equipment is configured in a shippable box, there is the third dimension (the height) of this box that can be used to support additional equipment above grade (since the structural steel is already there). As a result, the PAU equipment layout will not look exactly like the stick-built equipment layout, and the opportunity to use the height should not be lost when evaluating options on the process layout of equipment.

Again, a key takeaway on the PAU is that IT WILL NOT look like the equivalent stick-built equipment layout. That being said, the PAU basically contains an entire portion of a plant. It may be a sub-unit, unit, or even an entire process. The split is usually determined by the ability to include all the equipment associated with that portion of the plant along with the piping, electrical, and instrumentation to make that PAU more or less complete. The reason for trying to get an entire unit or part of a process within the single PAU is that one now can also include the electrical and instrumentation hook-up, pipe supports, control valve instrumentation, and all the cabling for the operation as well as low voltage power in the PAU during construction offsite.

In some of the larger PAUs, including a motor control center (MCC) or remote terminal unit (RTU) within the PAU structure is possible. Again, such an arrangement provides the opportunity to install, connect, and test many of these components off the site at the fab yard, removing additional labor from the project site. Everything is contained within the structural framework of the module—equipment, piping, pipe supports, valves, instruments, electrical components, and cabling. Additionally, as with the PARs, painting, heat tracing, insulation, and fireproofing are applied as far as is practical.

Piping supports will be incorporated within the larger PAUs, depending on the equipment layout and needs. In some cases, these pipe supports can be as large as small independent PARs and incorporated along the outside edge of the PAU, combining both into a single unit and permitting more internal piping hook-ups (PAU to

PAR), leaving only a few main connections to be made between the modules when they arrive at the site.

4.1.3 Skid Mounted Unit (SMU) or Vendor Package Unit (VPU)

A skid mounted unit (SMU) or vendor package unit (VPU) is commonly referred to as "equipment skid" or "equipment package." This is where the definitions begin to deviate depending on who you are talking to. The differentiator is that the SMU or VPU is typically a piece of equipment/process that has been more or less mass-produced and built on a skid to facilitate shipping. The SMU is typically designed and put together by a single "vendor." This vendor controls all aspects of the design, procurement, fabrication, and assembly of this package. It contains components that the vendor will be engineering, designing, fabricating, and assembling as a module.

This SMU or VPU can be a direct off-the-shelf item, but typically these units end up having enough variability that they must be adjusted by the vendor with small variations to the equipment, piping, electrical, controls, and instrumentation. They are completely piped and electrically wired to skid edge and complete with access platforms, ladders, junction boxes, and electrical connections. However, for all the components to end up on a transportable frame, the vendor has typically pre-identified and developed one or more common skid sizes as a basis for design. Examples of SMUs include anything from simple mixing equipment, heater units, air compressor packages, all the way up to larger units combining rotating equipment drivers and compressors with their receivers or associated interstage dehydration and cooling equipment. Other examples are some of the larger MCCs, fired heaters, and boilers. So, the SMU or VPU is really a very small version of the PAU that is built to be complete and functional when shipped. Basically, ready to set, plug in, and turn on.

4.1.4 Vendor Assembled Unit (VAU)

A vendor assembled unit (VAU) is commonly called just that, a "vendor assembled unit." It is an SMU or VPU that has so much more equipment from other third party companies and manufacturers that the VAU vendor must actually become a small engineering, procurement, and construction (EPC) contractor in terms of design, coordination of procurement, and work with third parties

that make the final VPU a much more unique piece of equipment.

We make this a separate module designation because this type of vendor has developed the additional complexity within their organization to perform all these additional "EPC-ish" activities. These require additional personnel for the much more complex design efforts and separate groups on the administrative end to interface with the client and their requirements and relay these to the manufacturing and assembly teams within the vendor's organization. In summary, this vendor has bridged the gap from a commodity skid supplier to a unique assembly EPC company.

4.1.5 Pre-Assembled Structure (PAS)

A pre-assembled structure (PAS) is commonly called a "structural module." This is the last module type of module terminology that we consider commonly used. It is usually a module built entirely out of structural steel for the purposes of support or access to parts of other buildings or modules. Common examples include the stairway access platform units (and sometimes even elevator units) that are scabbed on the outside edges of a PAR or PAU.

A PAS is built as a separate unit, allowing it to be shipped separately on a much smaller transport or vessel and stacked or otherwise arranged to minimize the transportation footprint of the larger PAU it will be attached to. Also, because these units are typically identical, these PAS units can be fabricated and assembled in a separate part of the fabrication yard, using more of a manufacturing technique for their assembly line construction. They can be designed to be stacked or packed for shipping, taking up much less space on the transport.

4.2 Other Terms

Beyond these major module definitions is a vast range of other terminology to further dissect or describe subsets in terms of their size, method of transport or setting, or function. These descriptions vary from area to area and project to project, so it is important only to identify that they exist and briefly describe why one or more might pop up in conversations.

4.2.1 Small PAU/Medium PAU/Large PAU/Super PAU/Mega (or Small/Medium/Large/Very Large/Mega Equipment Modules)

Some prefer to further dissect the definition based on the relative size of the final module. For the PAU, they may typically be further divided up, as noted above. The problem with such a set of definitions like these is that everyone has a different concept of what "small," "medium," "large," and even "very large" is. Still, many feel a need to further define modules by size to give the reader an idea of the mix of modules in a project. But, additional information on the mix of sizes can communicate better information for the planning in terms of fabrication, transportation, setting, and hook-up. For example, a project consisting of 4 or 5 large PARs along with 8–10 large PAUs, is very different in configuration than that same project composed of 20–25 medium to small PARs and 15–20 medium PAUs. The smaller modules are easier and faster to build and are impacted less by missing or late equipment than their larger counterparts. They may be able to be lifted into position rather than required to be rolled onto their foundations. However, this flexibility is offset by an increase in on-site hook-up and precommissioning that must now be handled at the site rather than in the fab yard.

So, what are the typical split points for these very generic and subjective size ranges? We prefer to split them in terms of their transport and setting restrictions or characteristics. Table 4.1 shows the typical size and weight ranges by module type. A detailed explanation of them can be found below.

4.2.1.1 Truckable PAU

A truckable PAU is an equipment module that has been limited in size to fit within the weight and height restriction of a country's highway system (see Figure 4.1). For the US, this means the PAU is limited to an overall size of 14 ft by 12 ft by 80 ft long and a weight of about 50 tons. These dimensions can be increased to special "permit" load designations where the specific route has been analyzed, and State and Federal movement permits have been approved to utilize special load distribution trailers. Size and weight for this module type are driven primarily by the height limitations for bridges, the width limitations dictated by the roads or highway lane configurations, and the load distribution limitations on the highway pavement.

Table 4.1 Typical size and weight ranges by equipment module type.

Equipment module type	Size (ft)	Weight (short/US ton)
Truckable PAU (Equipment Module)	14 ft by 12 ft by 80 ft	About 50 tons
Small PAU (Equipment Module)	12–24 ft wide by 15–20 ft high, and 80–120 ft long	About 80 tons
Medium PAU (Equipment Module)	24–50 ft wide, 30–40 ft high, and 75 ft up to a max of about 125 ft long	80 short tons to up to about 250 short tons
Large PAU (Equipment Module)	70 ft wide and 125 ft long but up to 100 ft high (or more)	300–400 tons all the way up to 1,500 tons
Very Large PAU (Equipment Module)	120 ft wide by 240 ft long by 120 ft high	1,500–6,500 US tons (or more in special circumstances)
Mega PAU (Equipment Module)	140 ft wide by 100 ft high by up to 460 ft long	8,000–12,000 tons

4.2.1.2 Small PAU

An equipment module that is larger than the truckable PAU in terms of width, height, and length must be moved by an escorted special transport vehicle (see Figure 4.2). These are typically multi-wheel lowboys, some with steerable rear assemblies for maneuvering around corners and tight turns. Tractor and trailer configurations vary

Figure 4.1 Condensate stabilizer shipped horizontally as a truckable module. Source: Courtesy of S&B Fabrication Services.

Figure 4.2 Truckable module with permit load on specialized trailer. Source: Courtesy of S&B Fabrication Services.

Figure 4.3 Small pre-assembled unit being transported by 64-wheel rear-steerable lowboy with push tractor.

from company to company and trailer manufacturer, but all have the same goal in mind—provide a transport vehicle and trailer with enough wheels and axles to limit individual weight per axle to something acceptable to both the State and Federal road standards. The small PAU can be considered anything in the range of 12–24 ft wide by 15–20 ft high and 80–120 ft long. The module shown in Figure 4.2 was an oversize module requiring a special highway permit and a specialized trailer.

The small PAU shown in Figure 4.3 weighed approximately 80 short tons being moved with a 64-wheel trailer configuration and "push" tractor to overcome some of the mountainous grades in Idaho and Wyoming.

4.2.1.3 Medium PAU

A medium PAU is an equipment module that can accommodate a greater maximum height by careful route selection, avoiding standard overpasses as well as lower hanging electrical and power lines of the area. A wider module width is realized via careful pre-planning and coordination with local and State highway authorities in order to obtain special permitting, allowing the module to be transported by using the entire roadway width (with the appropriate escort vehicles involved).

These medium PAUs are typically in the range of 24–50 ft wide and 30–40 ft tall, with their length varying from 75 ft up to a max of about 125 ft, primarily driven by a function

of the road curve radius and corners they must negotiate. The weight ranges for these medium PAUs can be as light as 80 short tons to up to about 250 short tons.

The medium PAU may require some re-routing of cable and secondary power lines as well as potentially some temporary road shoulder stabilization work in tight curves. Generally, the module is sized to minimize the impact (cost and permitting) requirements for making these temporary modifications. These modules' weights can range from 80–250 short tons or more and typically require special multi-axle trailers that may or may not be self-propelled to move the module while properly distributing the weight.

4.2.1.4 Large PAU

A large PAU is an equipment module that must use a specially built heavy haul road or route as well as multiple lines of self-propelled modular transporters (SPMTs) (Figures 4.4 and 4.5). This module will be similar in size in terms of the maximum width and length as the medium PAU, but the height has been increased to about 100 ft. This extra height allows for much larger vessels and equipment to be installed and hooked up within the module. In many cases, the taller process towers may even extend above the top supporting structural steel.

It is this increase in height that provides all the challenges in terms of transport, both from a vessel sea shipping

Figure 4.4 Completed large pre-assembled units being moved to roll on–roll off (RO-RO) vessel for load-out.

Figure 4.5 Large pre-assembled units being prepped for load-out on multiple self-propelled module trailers.

standpoint as well as a land-based transport, and this is why the large PAU is defined as a separate case. Modules of this height can be accommodated on many of the same roads as their smaller counterparts simply by adding more axles of support—thereby spreading the heavier load over a larger area of the roadbed. However, the height is such that these modules will not be able to get under major transmission power or over conventional bridges. Re-routing overhead transmission lines is an option, but typically the expense, long lead planning, and other mitigation efforts make such an option a non-starter. But, with planning, such re-routing or burial of these overhead transmissions lines has successfully occurred. So the large PAU transport is limited to routes that can avoid these height restrictions both on land and even up navigable waterways. This makes most applications of such large modules limited to a project site on the coastline or near the coastline. The typical range in size for these modules is 70 ft wide and 125 ft long but up to 100 ft high (or more) with a weight from 300–400 tons all the way up to 1,500 tons.

4.2.1.5 Very Large PAU

This equipment module range covers all land-based sizes above 1,500 tons. These are the King Kong or Godzilla of the land-based module spectrum. Typical very large PAUs can run as heavy as 6,500 tons and are typically used where extremely large equipment must be combined on a single transportable frame in order to get the

advantages of being able to combine all the equipment of a subsystem into a single module, as well as the power, electrical, and instrumentation needed to make this subsystem or unit fully functional, or at least as close to ready to "plug and play" once it gets to the project site.

Remember, the concept of modularization is the same in all of these cases. The goal is to move as much of the construction, hook-up, and pre-commissioning off the project site (where craft labor is expensive) to a fabrication yard (where craft labor is cheaper). To be able to include the electrical, instrumentation, and control systems (EI&C), all the equipment making up that sub-process must reside on that one module. So, obviously, if the equipment is very large, the overall module size, weight, and footprint will need to be very large to accommodate it.

In the case of these very large PAUs, the process system it supports is made up of very large equipment and machinery, requiring that the module be large enough to accommodate it all. But, along with this increase in dimensions due to the large equipment, come all the challenges with transport and setting. Large seagoing transport vessels must be contracted. Not so much for their capacity or length, but to accommodate the width of these modules. On the receiving end of the journey, special docks that can handle the width and weight of these modules must be available or specially built.

The road to the project site is no longer simply called a "road" but has attained the status of "heavy haul route" with its increased width and special-purpose load-bearing across this width as well as, in some cases, an unrestricted height element.

Finally, there is realistically only one method of setting at the site, and that is to drive the numerous lines of SPMTs between the pre-set pier footings and then lower the massive module down onto them with the hydraulics of the vertically adjustable SPMTs.

All this makes the very large PAU a very specialized, complex, and more expensive option without even considering the need for the utilization of tubular columns rather than conventional structural steel I and W shapes. In fact, these very large PAUs begin to resemble some of their big brother offshore counterparts.

Figure 4.6 On-loading of a 4,500-metric ton module. Source: Courtesy of Mike Webb.

Table 4.2 Typical size and weight ranges by pipe rack module type.

Equipment module type	Size (ft)	Weight (short/US ton)
Small (or truckable) PAR	15–24 ft wide x 13–29 ft high and length up to 100 ft	Up to 50 tons
Medium PAR	35 ft wide by 35 ft high (without legs) and 130 ft long	50–400 tons
Large PAR	35– 55 ft wide, 75 ft high (without legs) and 120–170 ft long	500–1,200 tons
Very large PAR	55–120 ft wide, 75–100 ft high, and 170 ft to about 200 ft long	1,500–3,000 tons

So, why do they exist as an option? As mentioned, they are especially suited for modularizing very large equipment in project locations where the labor (cost and availability or both) is not there. With that introduction and explanation, these modules are differentiated by their width, length, and height. They can range from 1,500 to 6,500 US tons (or more in special circumstances) with dimensions of 120 ft wide by 240 ft long by 120 ft high (see Figure 4.6).

4.2.1.6 Mega PAU

As an aside, the authors have performed feasibility studies where mega PAUs have exceeded the dimensions and weights of even the very large PAUs discussed above. In these special situations of scarce craft labor coupled with a short construction window (in the northern regions of the world), these studies have further limited on-site work to the absolute minimum, produced mega PAUs where the entire process train might be on a very few modules. One such example study configured an entire plant on four combination PAU–PAR configurations. These super PAUs were estimated at approximately 140 ft wide by 100 ft high by up to 460 ft long with a range in weight from 8,000 to 12,000 tons.

4.2.2 Small PAR/Medium PAR/Large/Very Large PAR (or Small/Medium/Large/Very Large Pipe Rack Modules)

The train of thought used in the above PAU is similarly followed for the pipe rack modules. These are divided in terms of size as well as composition (see Table 4.2) in an attempt to further differentiate them.

4.2.2.1 Small (or Truckable) PAR

This is the most common (and most commonly seen) module pipe rack. One can drive down the highway and see its identical brother, the stick-built pipe rack in any plant or process facility. The small PAR consists entirely of pipe being supported in a standard frame and is very similar in composition to its counterpart — the stick-built pipe rack. In fact, if you do not examine either closely, you would be hard-pressed to determine if the pipe rack was stick-built on site or pre-assembled and brought in.

The difficulty in visually determining if the small PAR was stick-built or modular is because the structural steel configuration of both is almost identical. The pipe rack bent (the individual "A" frame supporting the pipes) spacing is similar, as it is based on the ability to properly support the pipe span. Typically for everything but a very small diameter pipe, this spacing is set at 20 ft. For the very small diameter piping (such as 2 inches and 3 inches), an intermediate support is provided at 10 ft spacing. Vertical spacing between layers of piping is also consistent between the stick-built and modular pipe racks at about 6–8 ft. The height is typically chosen to allow enough room to "roll" the pipe up on a 45 degree angle for branch connections on and off the main runs as well as loops for expansion. The difference is in the structural steel supports required on the bottom of the pipe rack, depending on if it is to be rolled into place or in the strength of the column steel if it is to be vertically lifted and set in place.

Since the pipe rack is designed for pipes full of liquid, but it is shipped empty, there is usually enough safety factor with the empty shipping configuration to accommodate any transportation accelerations. In addition, along the US Gulf Coast site (USGC), the requirements of hurricane and seismic considerations typically are more severe than the shipping forces. So, if these pipe racks are properly designed for the in-situ conditions, there may be no extra steel required for the modular design.

4.2.2.2 Medium PAR

This is a small or truckable PAR that is too big to be "truckable." Perhaps a play on words, but just like the equipment PAUs, when a PAR becomes too big, it demands a different mode of transport, typically moving from the trailer or multi-wheeled lowboy to the SPMTs.

These medium PARs can range in size from small sleeper (single-level) racks and truckable racks to multilevel pipe racks carrying the majority of the large process and utility piping down the center of a complex, and running 35 ft wide by 35 ft high (without legs) and 130 ft long or more. The benefit of the longer pipe rack is the ability to connect multiple 20 ft sections of pipe at the fab yard assembly in order to reduce welding connections at the site.

Since they are not extremely heavy or large, in many cases, they do not need specifically dedicated RO-RO vessels but can be transported via lift-on-lift off (LO-LO) self-geared ocean-going vessels.

4.2.2.3 Large PAR

The differentiator in this category is the aerial coolers (also called fin fan coolers) which are frequently installed on top of these pipe racks as they provide a great place for locating these induced or forced air draft coolers. At this higher elevation, they are able to move cooler air volumes through the coolers well above the equipment, where the resulting hotter air can dissipate without impacting temperatures around the equipment.

This large PAR deviates from the small and medium PARs in that it does require more structural steel to support both the cooler weights (which can be up to about 1/3 of the total weight) as well as the sometimes-extensive grated walkway under and around the coolers

Figure 4.7 Large PAR being fabricated end to end at the fabrication yard.

for maintenance access to the coolers and motors. In addition, there is more weight potentially added for a fireproofing coating on the primary support members where the heat from an ignited leak of these coolers' contents might compromise the structural integrity of the entire structure. These factors make the structural steel design a bit more complex than the small and medium PARs.

Because these large PARs may be designed to support the banks of coolers, they are typically much wider than their small and medium PAR counterparts. Their widths can vary from 35 to 55 ft wide, with lengths as long as required to provide support for the entire bank of coolers manifolded together (from 120 to 170 ft), with a height of about 75 ft (without legs) and weight from 500 to 1,200 short tons. Figure 4.7 shows a large PAR being fabricated at the yard (in line with several others) and later being moved via SPMTs into position for load-out (Figure 4.8).

4.2.2.4 Very Large PAR

This is a variation of the large PAR but with a much wider set of coolers, typically stacked two wide. Because of the size and width, as well as the installation of a complete grating walk area (sometimes called a "dance floor") under these coolers along with the associated access stairway structures, the structural steel requirements and piping arrangement require an even more complex design and support effort.

Figure 4.8 Large PAR being moved into position for load-out by SPMTs.

These very large PARs are much wider to support the double banks of coolers. Since the goal is maintaining these coolers in the general vicinity of the equipment, the width more than doubles to support this cooler configuration, increasing from the typical 55 ft average of the large PAR to a 120 ft maximum. The length may also increase, but only slightly, going from the 170 ft of a large PAR to about 200 ft, depending on the number of coolers in the associated bank. The height may increase slightly due to the increased complexity and size of piping and the need to potentially add an additional level for non-associated piping, approaching 100 ft tall. All these increases result in the weight of the very large PAR approaching 1,500–3,000 short tons.

4.3 Module Considerations

We hinted at some of these in our explanations of the various types of modules, but let's take a more structured look at some dos and don'ts.

4.3.1 Plot Plan Development

Earlier in our module definitions, we cautioned about falling into the design/layout trap of drawing "boxes" around groups of equipment in a stick-built plant and calling each a module. The initial reason for doing this may go something like: "this (stick-built) plant is a second- or third-generation design and as such has been optimized as much as it can be." That may be true in the context of

the traditional two-dimensional stick-built layout, but remember, this stick-built layout did not optimize the areal extent of the plant's footprint—at least not to the level that it needs to be optimized for a modular plant. Usually, there was little need to—space was not a premium, and in many cases, the more distance between pieces of equipment, the better. As such, the stick-built plant was laid out to provide access, ease of maintenance, and in some cases, separation where distance created a safety buffer. This works well for a two-dimensional plot plan layout that can easily spread equipment out to create these maintenance areas and safety buffers.

But, for a modular project, using a stick-built plot plan as an arbitrary guide in the development of the module limits fails to take into consideration two very important benefits provided by the module:

- **the third dimension (available opportunity to "overlap" levels of piping and equipment within the module):** this has the potential to allow re-orientation of equipment—providing opportunities to elevate equipment that will naturally need to flow downward, potentially eliminating pumping as well as reducing the footprint of the equipment group—both of which ultimately reduce piping length.

- **the available steel structure:** this supports the equipment and provides access from the sides and from above this equipment and piping, creating additional opportunities in terms of maintenance access from above or either side.

4.3.2 What Does Such an Exercise Look Like?

Figure 4.9 shows a stylized petro-chemical process plot plan layout of the six sub-process areas as a stick-built effort. The six areas represent the separate sub-process systems that made up the plant. As you can see, because there were only two dimensions to lay out the equipment, it was difficult to get all the equipment situated in close proximity to the other equipment within that sub-process. Figure 4.10 was the same process plant when each of these six sub-process areas was "modularized" by combining like process components within the same module. Except for some rotating equipment that must be set on concrete foundations, all of the equipment has been optimized into these six modules. The typical pipe rack that runs like a backbone through the center of the stick-built

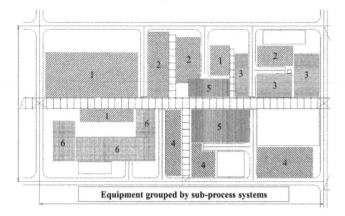

Figure 4.9 Stick-built plot plan showing outlines of the six sub-process areas.

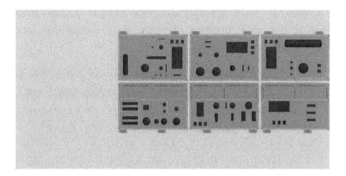

Figure 4.10 Equivalent module "6-pack" layout with each sub-process as a separate module.

plant still exists but now has been included as part of the three lower modules.

Note that these two plot plan areas are drawn to scale, so along with the movement of much of the same equipment into proximity to each other, the overall process footprint was reduced. Because the overall footprint has been reduced, there is a knock-on effect of reducing the overall piping lengths (assuming the piping/fitting ratio remains relatively the same). The result is an overall savings of piping materials.

In addition to the configuration of equipment and piping into sub-process-oriented modules, specific areas within the stick-built plant require varying degrees of craft support. Figure 4.11 is the same stick-built plant layout except that the plant equipment has been split into different areas depending on the craft work-hours required to complete and install pieces of equipment within that area of the plant. With such information on craft hours to assemble, set, and hook up individual pieces of equipment or areas,

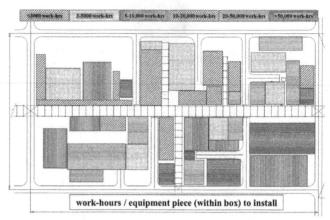

Figure 4.11 Same plot plan example but sorted by construction site craft work-hour density.

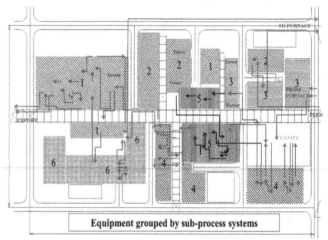

Figure 4.12 Process flow overlay of the stick-built plot plan.

this breakdown can be used as an additional guide on how to prioritize what equipment gets modularized and what can be left off for eventual setting at the site.

4.3.3 Why Analyze the Process?

In the typical refinery, chemical, or processing plant layout, the final product is developed from a set of linked multi-stage steps. That is, the process outputs from one sub-section feed the inputs to the next sub-section, making them somewhat linear in terms of layout and equipment configuration. However, for various reasons, the actual equipment layout for the operating stick-built plant may not reflect the simple and seamless linear layout that one might expect from a process block diagram or process flow diagram (PFD) review. (The many reasons for such discontinuities are beyond the scope of this discussion.)

Suffice to say, the result is that different pieces of equipment supporting their common sub-processes are split and scattered throughout the entire area of the plot plan. This means an equipment re-arrangement is required for effective modularization.

4.3.4 Steps in Maximizing Module Efficiency

Initial discussions should begin regarding how the entire process works and what equipment supports others within each sub-process. In every case, the first step in this analysis is the development of a process flow diagram overlay on the current operating plot plan (Figure 4.12).

This should be made as detailed as necessary to identify the various flows, where they come from and go to, major inputs, and product outputs. This becomes an "as-built" PFD re-oriented to reflect the process flows as they directionally occur in the operating plant. This provides the team a quick visual of the main process flows back and forth, the complexity of areas, and potential opportunities for consolidation of equipment to reduce nonessential back and forth piping runs.

Along with this visual process flow overlay on the operating plant is a separate exercise on this same plant with respect to the identification of all equipment supporting each of the sub-processes within the operating plant. If unfamiliar with the process, a good place to start is to look at how the PFDs are split up. They typically define and group equipment by subsystem and show the entire grouping on one or more sequential PFD pages. The pictorial example shown in Figure 4.12 is the stylized output of such a process grouping. With these two pictorial representations of the process, the small team of a modular piping layout lead, a process engineer, and potentially a couple of construction and operations-oriented folks can begin the development of an ideal plant layout using the flexibility and opportunities provided by the three-dimensional module envelope.

What is typically discovered during these initial discussions is that while there was some discontinuity, there is still a lot of continuity in terms of equipment arrangement and proximity. As such, it becomes relatively easy to highlight and show these equipment groupings as the

six differently outlined areas identified in Figure 4.9. Where there were these discontinuities, the challenge then becomes to understand the process well enough to begin evaluation of trade-offs for combining all equipment within a sub-process in a contiguous area versus the reason(s) why some of these equipment pieces were separated in the first place.

Working with the PFDs, plot plan, and major piping flows, this piping plot plan team is able to re-configure the original operating plant two-dimensional layout. By vertically stacking equipment above the equipment it directly served and including them in the proximity of their supporting taller towers, they are able to come up with the equivalent process in a modular configuration (previously shown in Figure 4.10).

The result is further developed by the module piping layout lead for a preliminary review by the entire project team. The aim of such a review is to get a consensus that the proposed layout will work: process engineers assess impact to process and products; mechanical engineers assess adjustment to pumps and other mechanical conveyance equipment now at elevation; structural engineers assess the support of equipment and adjustments to accommodate; O&M assess access and operability; safety engineers assess fire risk and ergonomics; project managers control schedule and cost implications, and so on. The result then becomes a true team effort in terms of the modular design.

Again, why all this effort to re-group and consolidate? Remember, if one of the goals of modularization is to move as many craft labor work-hours off the project site and into a fabrication yard, this can only be maximized when each module is designed so as much as possible of the equipment and processes within that module can be fabricated, assembled, tested, and pre-commissioned off the site in the fabrication yard. In many cases, like the example shown in Figures 4.13 and 4.14, the results can look extremely different.

Hopefully, this has provided an idea of a practical approach to initially set up and develop a modular layout from a stick-built plant process in order to obtain some of the benefits of modularization identified in Chapter 2. Details on how to evaluate these advantages will be further explained in Chapter 5, where more information is provided under the description of the business case for modularization.

4.3.5 Additional Module Considerations

While the two examples (Figures 4.13 and 4.14) are pretty impressive in terms of the considerable savings on the plot plan size and extensive use of the third dimension (height), there is much more to the exercise of modular development than simply "squishing" a two-dimensional equipment layout smaller and smaller until it goes vertical. Below are some additional considerations, in the approximate order they should be considered.

4.3.5.1 Maximum Module Size

One of the first questions to be resolved in a module project analysis is: What is the maximum size module that one can get to the project site? The maximum sizing is absolutely critical in the development of the most efficient modular plan. A preliminary transportation study should be kicked off very early in the project evaluation and development (even as early as opportunity framing or FEL-0) with the goal of determining what this maximum module sizing (or maximum shipping envelope) could be.

This is important in terms of how almost everything else is analyzed for the project. For example, it will determine how much of the equipment has the potential to be modularized, how easy or difficult it will be to include enough of the related equipment to make the E&I&C hook-up and pre-commissioning efforts off the site beneficial and, as a result, what the work split will be between the module yard and the project site.

4.3.5.2 Optimum Module Size

Approaching the module sizing from a different direction (via a process optimization) will provide the optimal module in terms of the equipment and configurations required to maintain the entirety of a sub-process or process within a module. However, when the piping layout team starts putting all the pieces together, they may find the resulting module is much bigger than the maximum shipping envelope will accommodate. This is where the give and take between the module team members (process/mechanical/ civil and structural/piping layout) begins in earnest.

Evaluation of the difference between the current maximum shipping envelope and the desired module size in terms of the process will kick off additional studies pertaining to a review of alternatives and their associated

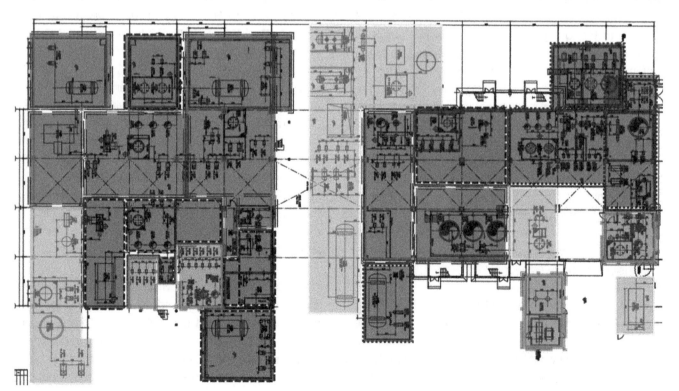

Figure 4.13 Originally proposed stick-built plant with different processes Source: Courtesy of KBR.

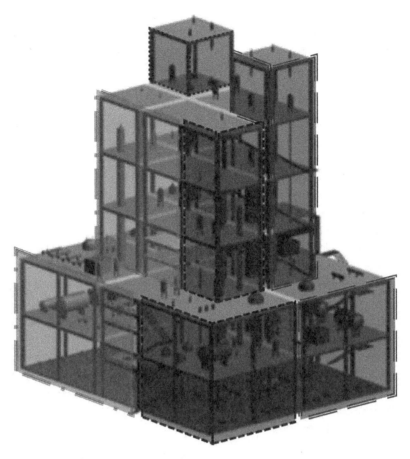

Figure 4.14 Equivalent four-module modularized plant. Source: Courtesy of KBR.

costs to modify one or the other. For example, the project may look at options to increase the clearances along the module travel route or, conversely, the impacts of the extra steel and labor to further divide the desired process modules into something smaller and the resulting impact in terms of additional craft labor and schedule time at the site to re-assemble. There will always be an optimal maximum size for a given project and location.

4.3.5.3 In or Out of a Module?

As the modules get smaller, either due to shipping restraints, fabrication yard limitations, or site setting restrictions, more and more "large" equipment will need to be left out. However, this is not the only reason that one should make a critical analysis of what goes in a module versus what comes out. Regardless of the size and shape of your module program, there should always be a critical evaluation of certain pieces of equipment to determine if it makes sense to "modularize" them. The

following are some of the obvious initial analyses that should be made.

Large/Heavy/Tall Vessels The first analysis should be on anything very large, tall and/or very heavy, with relatively few and simple connections needed for hookup. In many cases, while the actual piping on the vessel may be complex, the piping to and from the vessel may be simple and relatively easy to install. If the vessel can be "dressed" (that is pre-outfitted prior to shipping or outfitted off site) prior to setting, then the only "extra" work required of the site construction team is the connecting of the "in" and "out" process piping.

The reason for this analysis is that a large/heavy/tall vessel can be a problem from a structural steel design standpoint. These vessels can impose some significant moments on the base, which must be resisted by the module steel design. This makes the module structure more complex and increases the size of these support members. Besides taking up extra area on the module, all these additions add weight and cost to the module. Sometimes pouring a separate foundation adjacent to the module and setting the vessel separately on it at the site is a more economical choice.

Rotating Equipment   Large rotating equipment is another case that should be considered. Historically, this large equipment was required to be set above grade to position inlet and outlet compression piping below a compressor to avoid a deep excavation for this piping, as well as make it easier to service the compressor since all discharge piping is below it. This, coupled with the dynamic forces developed when the rotating equipment was running, required a massive elevated concrete foundation (sometimes referred to as a "tabletop"). The tabletop is designed as a massive concrete structure to provide the necessary mass to dampen the dynamic forces being developed by the rotating equipment to keep these forces (in terms of vibration) from being transmitted through the piping. Figure 4.15 is the inside of a completed compressor tabletop building and provides an idea of the magnitude of these massive structures and the equipment they support.

The alternative to this is, with enough engineering and design effort, this rotating equipment can be modularized. However, the design is more complex as the structural steel, which does not have the mass that the traditional concrete has, must be designed and configured properly to withstand and dampen any vibrations.

In the past, this option has been limited to the smaller capacity equipment and better suited to turbine and turbine derivative-type machines where the dynamic forces are not as severe. However, companies such as MHI are successfully modularizing larger and larger compressor equipment (Figure 4.16). Where such alternatives exist, an analysis of the costs (and project schedule time required) to perform the design and dynamic simulations for the module steel alternative must be "weighed" (no pun intended) against the costs and timing of pouring these massive above grade foundations at the site.

Another solution to the tabletop requirement is to use compressors with inlet and outlet nozzles facing up. This allows the compressor to sit at grade, with all piping connected to the top of it. The downside is that any compressor work will now require that all this large, heavy piping be removed before one can access the compressor shell for major maintenance. However, with the increased reliability and available run time between major work on some machines, this option may not be as demanding as one might imagine.

4.3.5.4 "Spacious" Equipment

In some cases, the equipment being considered may cover a lot of areas or be made up of very large "pieces." An example is the large furnace or "heater" that is typically used to provide heat/steam for petrochemical plants. This furnace is actually an assembly of several primary pieces of equipment, including the radiant section,

Figure 4.15 Completed tabletop with compressor and building installed. Source: Courtesy of MHI.

Figure 4.16 Structural steel modular alternative for compressors. Source: Courtesy of MHI.

convection boxes, inlet and outlet ducting, sulfur recovery unit (SRU), and exhaust stack. Typically, the large radiant heating section can be further split up if vertical restrictions become a problem. Additional sectioning beyond this becomes problematic as it creates much more assembly work at the construction site than benefits from the partial assembly.

Other examples of equipment best brought in as pieces and assembled at the site are belt conveyor equipment used in mining, pelletizers (e.g., sulfur and other pellets produced in fertilizer processes) often set up under large building roofs, and delayed coker units (for handling petroleum residual oil—thick bottoms from initial oil cracking process) where large vertical vessels sit on top of concrete structures so the produced coke can drop out the bottom via a "cutting" process and then be belt conveyed to large circular areas for storage.

In most cases, these are best shipped to the site to be assembled. If possible, trial fitting of the pre-assemblies makes sense as this ensures a proper fit and reduces work-hours required at the site.

4.3.5.5 Module Density and Mix

One interesting line of thought is related to the number, size, and/or type of modules in a process plant area. Because the method of setting the modules is limited to lifting into position (for the smaller ones) or rolling into position (for the larger ones), one can end up with a module setting nightmare if not coordinated with the construction team early in the module planning development.

There are several opposing trends that need to be considered when developing the module size and mix to ensure a path of construction that will be executable:

- **A larger module takes longer to fabricate.** This sounds like a "no-brainer" but can become very troublesome when working through a tight schedule.

- **A larger module is more apt to be delayed by its equipment.** The same basic information is required in terms of up-front design details of the equipment and piping to properly design the structural steel as a small module. However, with more equipment and interactions, the larger module has a greater chance of being impacted in some of this early design development by unforeseen delays of critical information on one or more pieces of equipment.

- **Larger modules require specific setting paths.** While the smaller module can be set in place by crane with some flexibility on the location of this crane, the larger module is limited to being driven into position. The complex matrix of heavy haul routes and supporting piers dictates the setting sequence and direction. This creates a lack of flexibility in terms of setting options and timing.

When you have a local project site that will receive a variety of module sizes with different setting methods, there needs to be a concentrated coordination effort with many of the project groups to ensure there is adequate time to allow the critical path module, typically one of the larger modules, to mature from a design, procurement, fabrication, assembly, and testing standpoint in a manner that will support the overall schedule of setting all in the area.

4.3.5.6 Who Does What?

Or, more properly stated, what is the optimal division of responsibility? This question does not have a simple answer and is not solved by a simple exercise. The best split depends on many factors, including the following:

- the complexity of the project or process (ratio of project purchased tag items versus fabrication yard purchased items);

- the sophistication of the module fabrication facility support systems (in terms of capabilities to purchase bulks and even fabricate piping and equipment);

- the competency of module fabrication facility (the level of module complexity that the fabrication yard is capable of executing);

- the comfort of the owner/EPC contractor in using fabrication yard subcontracting and purchasing capabilities.

The decision on how responsibility is to be split for engineering, purchase, fabrication, and even assembly, should be based on the strengths of the owner, EPC contractor, and module fabrication yard with the aim of trying to minimize interfaces in the module building process.

For example, a fabrication yard with ties to vessel manufacturing facilities, an internal pipe fabrication yard, and a joint working agreement with a deep water shipping company can potentially cut out several potentially delaying company and contractor interfaces, leaving the module fabrication facility with the sole responsibility to purchase bulk materials, fabricate pipes and structural steel, fabricate vessels, assemble and test, load out, ship, potentially offload, move to setting location, and set up the module. This removes many critical interfaces with the owner and EPC contractor and puts the entire scheduling and costing responsibility on the module fabricator.

Of course, this also places most of the responsibility for meeting the project cost and schedule on the module fabricator and limits their ability to come back to the owner and EPC contractor for delays and subsequent compensation. However, it also limits the owner and EPC contractor in terms of being able to provide assistance in expediting areas that are falling behind. The owner and EPC contractor are typically left with the only option available, that being providing additional or incremental funding to either get the module back on schedule or expedite it.

The division of responsibility needs to be fair and matched to the capabilities of the module fabricator, the strengths of the EPC contractor, and any desires of the owner. We will be addressing the contracting in a later chapter, but suffice it to say that the contracting (including the division of responsibility) needs to be fair and agreed to by everyone upfront in terms of clarity and transparency. It does the project no good if the module fabricator is treated unfairly or the contract does not allow everyone to make an acceptable profit.

Problems with consistencies in terms of payment, material delivery, and any other critical module manufacturing processes need to be addressed upfront and resolved quickly. With the module fabricator having possession of the module and the project equipment, they have the advantage. It is too easy to, in effect,

"hold the module hostage" as the shipping date gets closer, either by not completing all the required assembly and testing activities or just slowing down on them. This is not a usual occurrence, but it is one that is available to the fabricator regardless of the contractual terms and can be used to help coerce solutions to major inequities in terms of the differing contract understanding and execution.

The best approach is always to maintain a collaborative atmosphere and good working relationships throughout the module fabrication efforts to avoid getting into a potential situation where major assets are at risk of being held up at a fabrication yard halfway around the world waiting on contract differences to be worked out. There is not much "pushing" an owner or contractor can do at that point.

4.4 Summary

We have spent a lot of time taking the simple module "term" and subdividing it into all of its various permutations and explaining why these various size and weight versions exist. We hope you have seen that this degree of differentiation is important because each specific size of PAR and PAU comes with its own advantages and disadvantages. Without a good understanding of the various options, deciding how many and how big is much more difficult. With this basic size and complexity understanding, we worked through a simple module sizing exercise, starting with the first steps in the analysis and working through some of the options and trade-offs that need to be considered when trying to determine the optimal module size and composition. Finally, we offered some additional considerations in terms of how to further optimize the module solution, just in case you may have thought this simple exercise was all you would need to become a master at this sizing.

There is a lot more to learn. This is only Chapter 4. Hang on!

References

Construction Industry Institute (2013) *Industrial Modularization: Five Solution Elements*. Austin, TX: The University of Texas at Austin: Construction Industry Institute.

O'Connor, J.T., O'Brien, W.J., and Choi, J.O. (2013) *Industrial Modularization: How to Optimize; How to Maximize*. Austin, TX: The University of Texas at Austin: Construction Industry Institute.

Tatum, C.B., Vanegas, J.A. and Williams, J.M. (1987) Constructability Improvement Using Prefabrication, Pre-assembly, and Modularization. Austin, TX: The University of Texas at Austin: Construction Industry Institute. Available at: https://www.construction-institute.org/resources/knowledgebase/knowledge-areas/design-planning-optimization/topics/rt-003/pubs/sd-25.

chapter 5 The Business Case for Modularization

As mentioned in the Introduction, if you have been diligent in the review of the previous four chapters, you will have developed a sound understanding of what a module is, why modularization is important, why implementation will be a challenge, and the potential options in terms of module configuration. With this grounding in the basics, we can now go into a discussion of the business case. This chapter will explain what makes a project a good candidate for modularization, the factors to consider, the importance of modularization consideration timings, laying out the modular project execution planning steps by project phases, and expound on how to conduct a business case analysis with one of our tools.

The business case provides answers to the following questions:

- Is modularization viable for my project?

- If so, what is the optimal amount of modularization?

- What is the timing for the module program set-up?

- How does the business case fit within the project development?

Unfortunately, while the business case model provides an analysis of whether a project should be modularized and some ideas on the timing of it, it does not help lay out some of the practical implementation aspects needed to actually set up a module job, such as:

- How to set up a module program (Chapter 6).

- What needs to be adjusted in terms of project procedures (**Execution Plan Differences**—Chapter 6).

- What it takes to make the module program successful (**Critical Success Factors**—Chapter 7).

- What additional efforts can be made to improve when producing multiples (**Standardization**—Chapter 11).

5.1 Fundamentals of the Business Case

5.1.1 What Makes a "Good" Module Candidate?

A project is a good module candidate if the benefits of modularization can benefit the project.

It is intuitively obvious, but the key to defining a "good" project module candidate is to understand *everything* that a modular design and execution have to offer and compare these benefits to all the challenges that the proposed project must overcome to see if there is a match. Some of the benefits to a project are obvious, as the example below illustrates.

THE OBVIOUS EXAMPLE CASE

This project site, located in an area of the world that is extremely cold and inhospitable, needed additional oil-, gas-, and water-processing facilities to increase the oil field's gas handling facilities, boost oil reserves, and increase oil production. The remote location would not support a major on-site construction team, so the only option was to build and complete the facilities in a more temperate climate and ship the modules to the site. In addition, in order to get to the project site with any of the equipment and materials, it must access it via open water, and this method of delivery is limited to an 8-week period when the local access waterway was not frozen and ice-locked.

The answer to this dilemma was to design and build this as a modular facility, resulting in the design and production of 26 oil-, gas-, and water-processing modules, fabricated and shipped in two phases over a two-year duration (during the 8-week open water period). The construction phase of the project was set up in New Iberia, Louisiana, where the EPC company constructed the 26 modules totaling approximately 36,700 tons and successfully shipped from Louisiana, down through the Panama Canal, up the West Coast, through the Beaufort Sea (during the 8 weeks the sea was not ice-locked) to the project site in Prudhoe Bay, Alaska.

The details of this very successful project can be found by Googling the Fluor GHX-2 project for Arco. The project was executed in 1993–1994, and the modules were the largest of their kind ever shipped at that time (with the largest at 208 ft long x 93 ft wide x 122 ft high with individual weights of up to approximately 5,600 tons). The project used many cost- and schedule-saving techniques along with this offsite assembly, including electronic badging, barcoding of materials and tools, high-volume flex core welding, assembly of pipe spools and equipment at ground level prior to installation, and extensive use of manlifts instead of scaffolding. In the end, the project was executed ahead of schedule and under budget and won several awards for its overall execution.

Unfortunately, the benefits from using an alternative project execution approach to stick building (like modularization) for most projects are not as intuitively obvious or recognizable. It is for all these other projects that this chapter on business drivers has been written.

5.1.2 Every Project Has Some Amount of Modularization

We cannot think of a single project that we have worked on that would not have benefitted from at least some amount of modularization (or pre-assembly). In fact, we would go further out on a limb and say that every project you as the reader have worked on or will work on will have some amount of "modularization."

Skeptical? Then consider the following—every plant is supported by utilities, including power generation, instrument air, nitrogen, or other gas production units, potable or non-potable water systems, water treatment, and solids disposal. All of these are processes that are made up of one or more skid units that have been pre-assembled off the site.

Why? Because most of these supplying companies are associated with the development and sale of a standard product or suite of products, or they have a product that they sell that has provided them the opportunity to design one and build many or, some other permutation of duplication, that made it profitable to provide a product that had some pre-engineering, pre-assembly, and testing. Even with very basic elements, like the typical American Petroleum Institute (API) or the American National Standards Institute (ANSI) pump, some modularization (pre-assembly) is involved. The pump and motor are set and shipped on their own stand or base with preliminary alignment pre-performed to make the installation of the pump easier and quicker. So, all the pre-assemblies mentioned above are in the literal sense of the term—small modules.

Of course, we are not proposing that any project with a pump in it be called a "modular" project. And, we are not asking you to consider the little pump skid in the same context as a 2,500 ton fully integrated process module. But we are suggesting that the same line of thinking used to develop the pump skid and the air compressor package be followed to expand the project thought process to include the much more complex set of equipment and piping for the purposes of moving from the traditional stick-built execution to a modular execution.

Yes, the above opening thoughts also have a lot to do with "Standardization," which will be addressed in Chapter 11. But for now, be open to considering the comparison in thought process between the simple skid package development and the much more complex integrated module solution.

5.1.3 Three Distinct Levels of Involvement

As we briefly explained in Chapter 1, Section 1.4 Three Distinct Module Options (or Circumstances), all module jobs can be sorted into one of the three options below.

1. **Minimal.** This level of modularization involves implementing only the very obvious beneficial options, for example, the typical pipe rack and any equipment skid efforts already suggested by the vendor. The inclusion of this level of modularization is easy and requires a slight deviation from the standard stick-built approach. It has little impact on the schedule because the only difference between the stick-built and module pipe rack is where it is being assembled. In many cases, the design can even be the same with some simple additions of base supports for shipping aids. However, this level of modularization offers few benefits in terms of cost or schedule savings.

2. **Selective.** This level of modularization covers everything from minimal to maximum and therefore is the largest category. It is determined by the critical use of the business case analysis presented in this chapter.

3. **Maximum.** This level involves the decision to modularize as much as possible due to project constraints, such as weather, location, lack of infrastructure, high local labor cost, lack of local labor, etc. (as noted in the example at the beginning of this chapter). It is a "no-brainer" in terms of the actual decision to maximize but very difficult in terms of identifying exactly what the optimized maximum option looks like. This is difficult because the obvious "maximized" option is to build the entire process plant as a single huge module. That way, it could be pre-tested and commissioned, and it only requires a simple hook-up to the input feedstocks and the output delivery methods for the product. Unfortunately, such a huge single module would be extremely costly to build in terms

of just the structural support steel alone, even if you could somehow find enough self-propelled modular transporters (SPMTs) with enough axles to move it and a vessel big enough to ship it.

5.1.4 Advantages and Challenges

In Chapter 2, we fully described the advantages and challenges of modularization in detail. They could be quickly summarized here, or what we think is a better option, we suggest you go back and review them yourself! But, after a bit of self-reflection on the "cheeky" suggestion of re-reading provided above, and thinking you might feel a bit abandoned, we came up with another way to describe and summarize both that may be a bit more memorable and came up with the following advantages and challenges "sound bites":

1. **Advantages.** Modularization removes scope from the site to be executed in a fabrication facility with a cheaper labor force that is more efficient and possibly more qualified, resulting in cost and schedule benefits to the overall project.

2. **Challenges (or Disadvantages).** How to successfully identify and provide the needed technical details and materials required to support the module fabricator's schedule from their first steel cut to their final module load-out for shipping. (Specifically, timely design development, material procurement, bulk purchasing, structural steel and piping pre-fabrication, spool and equipment delivery, structural steel assembly, piping and equipment installation, check out, pre-commissioning, load-out, and shipping.)

5.2 Important Factors to Consider

Identifying what needs to be considered when evaluating a project in terms of whether it should be modularized and then how much modularization should be performed on the project can be a complex and potentially convoluted process without a systematic approach. So, what we decided to provide is a typical approach. This is not the only way to approach this effort and definitely not a "one size fits all" in terms of suitability, but it will give you an idea of how to approach such a "business" analysis.

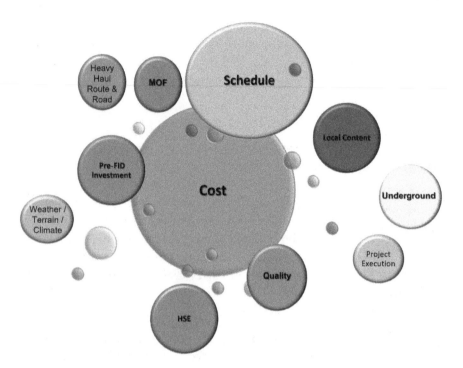

Figure 5.1 Project drivers and their relationships.

5.2.1 Universal Key Project Drivers

We have called this section "Universal" project drivers because many of these will have to be addressed in some form or fashion on every project, so it is vital that the project understands its implications to best accommodate them. Figure 5.1 provides some obvious clues regarding these universal project drivers. The size of the bubble in Figure 5.1 is proportional to the importance that the owner/client and the EPC contractor place on that particular universal project driver. The bold/normal font indicates who was primarily responsible for executing that driver, with bold being the owner/client and normal being the EPC contractor. These key project drivers will be explained below in the order of how these drivers happened to be systematically addressed in an opening discussion between the owner/client and the EPC contractor regarding a project. So, how should this cryptic bubble listing in Figure 5.1 be interpreted?

5.2.1.1 Cost & Schedule

These are #1 and #2 or perhaps #2 and #1. In either case, they are difficult to separate and need to be discussed together. However, for the typical owner/client, while they are also very important, they are usually considered independent of each other. That is, they see no reason why they cannot work to reduce cost and

then turn right around and incorporate all the recently agreed cost-reducing efforts into a new schedule that also shows a significant reduction. The problem with this approach is that the cost reductions being made have a negative schedule impact associated with them in many instances.

For example, purchasing a piece of equipment from the lowest cost vendor returning the bid solicitation package could provide a substantial cost saving. However, if this vendor is extremely busy (probably because they are so inexpensive), their proposed schedule for completing and shipping this equipment may be several months longer. Maybe not a critical issue on a stick-built job, but when that piece of equipment is part of a highly integrated module, that delay in delivery may force a hold on the module fabrication sequence (or at least complicate the assembly, causing extra work or re-work) until this piece can be received and installed.

There needs to be a serious discussion about both as a single topic. Specifically, which one—cost or schedule—is more critical? Because of the mindset noted above, this may not be easy for the owner/client who does not fully understand the interrelationships of cost versus schedule in terms of module fabrication sequencing. Since the approach to a module job in terms of sizing and scope of work splits is impacted by which of these two project drivers take a higher priority, it may be beneficial to overemphasize the magnitude of each when initially discussing to help everyone visualize and understand the situation.

So, it is suggested that the conversation center around a couple of assumptions that reflect a not-so-realistic, wildly high-cost delta along with a similarly huge schedule delta to make the point and get a decision made.

For example, what is more acceptable, saving $100M or reducing the overall schedule by 6 months? The key here is to make the delta differences very significant, but still within the realm of plausibility for that particular project (and owner/client) so that they are forced to consider the two options in their project net present value (NPV) analysis, which the EPC contractor never has access to or sees.

The costs deltas must be high enough to interest the owner/client's capital project managers. The schedule deltas must cut enough time off the end completion date to force a re-evaluation of early product delivery and its impact on the project NPV. For some projects, such as oil production, the answer is obvious—in the depletion analysis, any production volume lost by a later scheduled start does not get produced until the end of the reservoir life (at a very highly discounted present value rate). Production of a manufactured product is not as easy to analyze, but the owner/client has it in their ability to perform these alternative options on NPV and rate of return (ROR). But, once this determination has been made, the module design and layout team can consider different configuration options that will serve to meet one or the other a bit better.

5.2.1.2 Project Execution

We are skipping project execution for the moment as a project driver, because the EPC contractor is probably more interested in it and it is probably more in their control, so we move on to pre-FID investment. Check Chapter 6 for details on the project execution.

5.2.1.3 Pre-FID (Financial Investment Decision) Investment

This is indicated (by the size of its bubble in Figure 5.1) as an intermediate project driver. This is because it can

quickly become a major stumbling block in some module jobs in terms of the EPC execution planning.

In many cases, this is because most, if not all, owner/client organizations follow some sort of project "stage-gate" project approval process where a certain amount of project detail is required to advance through each "gate." Since some of this stage-gate process is prior to FID (financial investment decision) or project "funding," these efforts may run on a limited capital budget. This means there is little money to advance some of the detailed design/procurement of vendor design info, etc. in time. This is especially true of complex equipment located within a module or electrical or instrumentation equipment (e.g., motor control centers or remote instrument buildings) that rely on data that is later in the design sequence.

In order to meet the schedule, this long lead equipment must be either:

- accelerated during the procurement process once FID has been made; or,

- purchased prior to FID (and risk ending up with equipment commitments and no sanctioned project).

However, there is a third option that provides the best of both worlds as a combination of both of the above:

- Identify the critical long-lead equipment where either design details will be required early to design the structural steel module framing or where the equipment itself will be late in delivery.

- For the equipment with a late delivery, work with the module fabricator to develop alternative assembly techniques that will allow the late arrival with minimal impact to the module fabricator. For the equipment that needs enough specific design details to properly design the structural module steel, work with the owner/client in terms of developing a purchasing strategy that will mature the equipment bid and procurement cycle to identify and award the equipment purchase, but limit liability for payment of the equipment to some clearly identified early design deliverables that are required for the module structural steel design.

This two-pronged approach limits the capital outlay required prior to FID and typically can be arranged by the owner/client if known early enough in the overall stage-gate approval process.

5.2.1.4 HS&E (Health, Safety, and Environment)

These elements have been lumped together because they have similar attributes:

- These statistics are all important to the owner/client (but a matter of company life and death for the EPC contractor).

- It is hard to equate how money spent translates into benefits received: Efforts typically are made to identify these benefits in terms of results: low or zero accidents, a healthier and happier workforce.

- But poorly executed HSE programs can be disastrous in terms of consequences when there is a general failure to properly support them.

- The EPC contractor is judged especially hard in terms of a less than stellar HSE historical performance. In some cases, poor results in terms of lost time incidents (LTIs) or industry fines, or no clear plan of recovery, are prerequisites to dismissal from consideration for the proposed project and even future work.

- As such, the EPC contractor bears the brunt of the results of a poor HSE record with the owner/client applying pressure in terms of potential withholding of future work or even cancelation of current work if failure is severe enough.

- It is up to the EPC contractor to "win" the owner/client over with an outstanding HSE record. One way to help this is to show their attempts to move work-hours from the relatively riskier stick-built activities to a more controlled fabrication effort at a designated module facility, where more work is performed by higher-skilled individuals in a more temperate climate (sometimes indoors) and working the majority of the time at grade.

- Of course, the owner/client will also have concerns to be addressed in terms of the environment, but these are typically trying to limit the temporary impacts of the construction workforce in the area of the proposed project plant site. Again, modularization reduces the required temporary construction areas (typically 2 to 3 times as large as the final plant site footprint).

- When evaluating these three project drivers, the EPC contractor must understand what parts of the HSE program for this project can improve community relations with the public, environmental impact on the immediate area, or perhaps simply covering a need to help the local community in terms of infrastructure upgrades or additions of services, such as a new fire station and equipment or extension to the local hospital, etc.

5.2.1.5 Quality

This is the third leg of the project stool, where the other two—"cost" and "schedule"—have been previously addressed. It is often joked: "cost/schedule/quality—pick any two." For example:

- You can have something fast and cheap, but the quality might suffer.

- You can get a high-quality product quickly, but it may cost you more.

- You can get something high-quality cheaply, but it may take a long time to get it.

The underlying assumption is you can get any two exactly how you want them but never all three. Unfortunately, this is not a joking manner when it comes up in the middle of the project execution as a result of misunderstood expectations the owner/client has of the EPC contractor, or vice versa, due to miscommunication of the actual details of the final product being delivered by the EPC contractor.

So, what does this have to do with the modularization (or the price of eggs)? The issue with all these various discussions around the project management triangle/the triple constraint/or whatever you call this three-legged stool is that these discussions come about because the project drivers were not properly identified and prioritized at the beginning of the project.

Just as we discussed in the cost and schedule section of the project drivers, there needs to be a determination of which one is more important; so we have a similar issue with quality. While everyone will agree that quality is important, how important is it in terms of its relationship with the other two—cost and schedule for your project—AND what are the appropriate tradeoffs in terms of the give and take among the three? At the end of the day (or, in this case, at the very beginning of the project conceptual discussions), there needs to be a consensus in terms of exactly how these three project drivers relate to each other. Not only that, but there also needs to be a promise that the relationship among these three in terms of off-setting importance needs to remain constant throughout the project. If there is a need to adjust the importance of one over the others, this should be met with an expectation from both sides that there will be some significant changes to the project outcome. Because changes mid-stream are especially dangerous and damaging in terms of both cost and schedule, especially when these changes are initiated during the module fabrication, it is important from a module fabrication project standpoint to make sure all project parties understand the implications of the "no changes" mandate early in the project development.

5.2.1.6 Local Content

As noted in Figure 5.1, this project driver seems to have a larger-than-life level of importance to the owner/client. That is to be expected, as this owner/client will need to "live" with the local area folks during the long-term duration of the plant operation. Getting off on the right footing with your potential "neighbors" is critical.

Of course, the first impressions in this relationship are made by the EPC contractor during construction at the project site. This is why the owner/client makes this relationship with the local area folks a priority through the EPC contractor. And what better way to help build a long-term positive relationship with the local folks than by providing as many as possible long-term, good-paying permanent jobs working at the finished facility? Of course, the actual number of permanent facility workers is usually a relatively small percentage of the folks in the area, so the owner/client pushes the EPC contractor to find ways to use "local content" to help bridge this gap during the EPC construction effort where there are many times the number of temporary construction jobs than the resulting permanent facility jobs.

This local content effort is related to the skills of the local population and they are often quick to pick up the heavy civil functions, like earthmoving, concrete formwork, and rod busting. Later, as the project progresses, there may be opportunities to use some of the local machinery as well

as the electrical and instrumentation talent. Because the EPC contractor has a much greater ability to influence initial local content, this can be a bargaining chip of the EPC contractor in their negotiations with the owner/client and should not be ignored when discussing this project driver.

5.2.1.7 Materials Offloading Facility (MOF)

The materials offloading facility (MOF) is typically considered a temporary construction facility for the purpose of docking vessels that need to unload bulk materials and prefabricated modules. However, it has been given numerous labels, so let's start with a very brief explanation of the terms often used interchangeably for a MOF—wharf, pier, quay, and jetty—in the hope that we can help you understand the subtle differences rather than be confused by them.

Port: A sheltered dip in the natural coastline that is large enough to allow vessels to come in (it could include a wharf, piers, docks, jetties, as well as warehouses, loading and unloading equipment, etc.).

Wharf: A constructed structure parallel to the shoreline with a deck supported by piers or piling with earth on one side and water on the other, allowing vessels to come up and dock. It may also have quays and piers extending from it as well as buildings for storage and ship servicing inland of it.

Quay: Quay is pronounced "Key." It is similar to a wharf but built on fill with a solid wall that extends from the elevated platform down into the bed of the water, thus creating a solid face at the coastline, altered via reinforcements to allow a vessel to float up and dock parallel to it and against it.

Pier: A piled column and beam type structure (typically wooden) that protrudes perpendicularly from the coastline into deeper water, allowing the deeper draft vessels to disembark passengers or cargo.

Jetty: A long narrow earthen or rock-filled structure that protects a coastline from currents or tides.

Dock (or dry dock): This has nothing to do with any of the above as it is a container, either floating or built in the ground, and used to keep a vessel inside it moored at a constant height or high and dry for repairs.

Figure 5.2 illustrates the MOF.

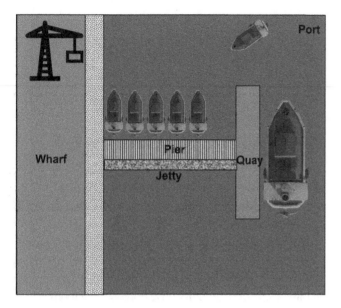

Figure 5.2 Depiction of the MOF.

Further variations between the terms:

Quay and jetty: A quay has earthen fill on one side and water on the other, a jetty also has fill, but the fill is internal to the jetty to allow water access on both sides.

Wharf and pier: Both are structural with piling, columns, and beams as opposed to earthen fill; the wharf is oriented between the water and land (and parallel to the coastline) while a pier has water on both sides and is typically perpendicular to the coastline.

Why do we bring all this up? Because your two authors have been misusing these terms "since forever," and it is important that we fess up and try to set the record straight on these terms.

Sorry for the digression, but it is important to know that this "ship to shore" connection, known by a bunch of different names, some correct and some totally misleading, has the potential to require a substantial investment consideration for the project where large modules need to be brought in from overseas via an ocean-going roll-on-roll-off (RO-RO) or a lift-on-lift-off (LO-LO) vessel.

Because of the weight of these larger module configurations, critical analyses are required of existing quays or wharves as well as the temporary MOF to ensure that a local failure of the quayside face does not occur. While this can be offset to a certain extent by using bridgework

to span from the vessel over the critical quayside edge, at the end of the day, the wharf, quay, or pier and its tie-backs into the bank need to be robust enough to withstand the localized loading of the SPMTs.

In addition to the actual cost and timing of building a project-suited MOF, in some areas of the world, there are all the permitting implications that go along with disturbing the coastline in such a grand manner. So, while it may seem like "small potatoes" in terms of total installed cost (TIC) (typical MOF costs can range from $20 million to $60 million), the larger concerns come from any permitting and environmental impacts that must be resolved. It is almost always more advantageous to use an existing facility.

In addition, there is the issue of the MOF or quayside depth in relation to the MOF deck height to contend with. The water depth at the quayside may become a limiting factor, with very large vessels not being able to ballast up (float) high enough to get adjacent to the quayside or vice versa, they may not be able to ballast down deep enough to line up with the top of the MOF deck.

5.2.1.8 Heavy Haul Route and Road and Weather/Terrain/Climate

Skipping heavy haul (HH) route and road as well as weather/terrain/climate as both being more in the control of the EPC contractor, we move to the final driver in this high-level analysis, Underground. Check Chapter 8 for more on this topic.

5.2.1.9 Underground

This typically is not addressed by the owner/client as a concern and is left up to the EPC contractor to figure out how they will install and route all the various underground piping and electrical connections. However, we identified it as an owner/client interest because underground issues are prevalent in brownfield projects, and the owner is always interested in how any new work will impact their current operations.

The use of modules in the construction execution plan can provide additional time (duration) from the initial construction site mobilization until the module arrival, thus providing additional time to perform this underground installation. This means that there will be more time on a project to perform underground surveys, and if unexpected lines or objects are found, more time to adjust or work around them.

5.2.2 Additional Project-Specific Factors

Besides the universal project drivers mentioned, there are many other project-specific factors that need to be considered when developing the optimal module solution. These are varied and widespread in terms of topic and impact. We have listed a few, only to provide an example of other factors involved in the module decision.

• project-specific scope

• site data

• module fabrication yard data.

Check Chapter 8 for more on these factors.

5.3 The Business Case Process

There is a process that every owner/client goes through to develop a project from its first idea or concept all the way through to the construction and start-up of the completed facility. The process has typically from five to seven phases, where there are certain deliverables that must be met at the end of each phase to move on to the next phase. These phases are defined by the level of engineering deliverables, the project estimate accuracy, or the project detail and management commitment. .

Every company has some sort of business case analysis tool. It may vary in the number of phases, the detailed requirements within each phase, specific project deliverables required, or may correspond to meeting an industry reference (e.g., Independent Project Analysis, IPA). It varies from one company to the next and is called by different names, but the goals are the same:

• to force the project to be evaluated in a consistent step by step;

• to make the project "pass" each phase before moving on and being evaluated in the next phase.

So, for example, for engineering, more detail is developed in terms of deliverables at each of these phases. Along with this additional detail comes the potential to produce higher accuracy in the overall project estimate. These additional details are included in a more defined project schedule. And the results of these efforts at each phase are presented to the company or corporate management (depending on the size of the project). The management will use this time to critically assess the project and details supporting it, ask questions, make comments, and in general provide a formal review (that should be not much more than a rubber stamp of all the previous informal interactions that the project management has had with company management—assuming project management was proactive in its efforts).

As mentioned, this analysis varies from company to company, but one consistent theme is that all companies have such a business case analysis procedure. Because there are so many versions of this business case analysis, it is important that everyone working with or discussing the attributes of a particular point in the business case analysis process must understand where the others are in the business case journey.

To support the reader's understanding of this process, benchmarking (Construction Industry Institute, 2012; O'Connor, O'Brien, and Choi, 2013) and (O'Brien, O'Connor, and Choi, 2015), the authors developed a Rosetta Stone on modularization business case phases (see Figure 5.3).

The top row in Figure 5.3 shows the seven project phases, which industrial companies commonly use:

• Opportunity Framing

• Assessment

• Selection

• Basic Design

• EPC (Engineering, Procurement, and Construction)

• Commissioning & Start-up

• Operation

Because some readers may not be as familiar with these seven phases, we have aligned the phases with the Frames of Reference (the IPA reference = the third row in Figure 5.3), Project Designation (EPC contractor reference = fourth row in Figure 5.3), and Cost Estimate Accuracy (the bottom row in Figure 5.3) (Construction Industry

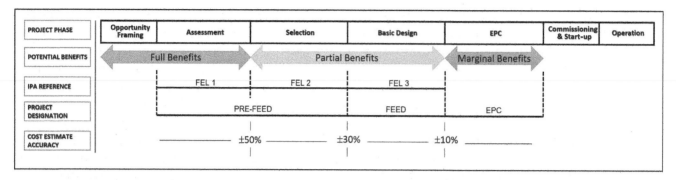

Figure 5.3 Rosetta Stone on modularization business case process phases.

Institute, 1986). This multi-level comparison approach is aimed to help someone figure out the maturity of a project by matching the definition for individual project maturity with one or more of these references and then moving vertically up to identify which of the seven phases that project was in. The reason for being so particular in this analysis matrix is that the proposed module business case analysis is going to be used at the end of each of these project phases. We bring this to your attention as this is the basis for much of our subsequent discussions and follow-up research work. It is vital that this basis be understood not only in terms of why it was set up this way but why it is essential to be analyzed in this manner. Finally, before we go into the details of the module business case tool (or flowchart), which will be explained in the subsequent section, we want to provide a sneak peek at why we are so intent on breaking the project up into these seven phases.

The second row of Figure 5.3, "Potential Benefits," illustrates why starting the module analysis early is so important. We will be going over the reasons in much greater detail, but if you can keep in mind that the optimal time for making decisions is during the first two phases of a project (Opportunity Framing and Assessment), the following explanations will make more sense. The large "band" of the Partial Benefits arrow that runs through the Selection (pre-FEED) and Basic Design (FEED) was drawn that way in order to try to include the entire range of module potentials. For the very complex module solutions, the time to make module decisions that will result in an optimal solution will run through Assessment and then start dropping off sharply in Selection (pre-FEED). This is due to the decisions made in Opportunity Framing and Assessment that impact global aspects of a project, such

as a project site location or the proximity to navigable water. Poor choices in location could eliminate the very large or very heavy module concepts before there is an opportunity to analyze them.

On the other hand, some projects may be able to still obtain maximum module benefits all the way through Selection (pre-FEED) or Basic Design (FEED). This is because the optimal module solution for that particular project is not sensitive to many of the typical decisions made during the project's first two or three phases. For example, suppose the process is such that the maximum module size can be held within the typical permit load for roadway transportation. In that case, the flexibility of site selection and logistics is greatly enhanced, almost to the point where it may not be critical. This is why such a broad "band" is shown in the Partial Benefits area.

Two points to remember in all this:

- One can NEVER look at the module pre-assembly options TOO EARLY!

- It is NEVER detrimental to the project to assume the module execution philosophy first and later fall back to the conventional stick-built approach if the module approach is not suitable for project execution.

We could spend an entire additional chapter on why these two points are true, but to put it simply, the earlier analyses required in the assumption that the project will be modular WILL identify many project issues at a much earlier date or phase of the project. This is because the systematic nature of the module analysis will force the owner/client to analyze similar project variables much earlier for the module job than would have been required for the stick-built job. In addition, by starting with the module option, the owner/client is forced to think about how to be creative in "packaging" all the equipment and piping into "boxes." This develops a collective project thinking where there is a natural optimization of equipment/ piping/process configuration that will inherently save money simply because the spacing between equipment is optimized and reduced (to fit within the module constraints). Should the project decide to be conventionally stick-built, this optimization can be maintained or relaxed slightly to accommodate a more spread-out plot plan approach. In either case, the final product in moving from the initial module concepts to a conventional stick-built approach is still probably more efficient than if it had started out as a stick-built philosophy. More on this to come in the later chapters.

5.4 The Business Case Model

Before you can set up a properly functioning module project, you need to determine if the project should be modular and the extent of the modularization. There are numerous ways to do this analysis. But what we provide makes the most sense (in our opinion) because it walks the analyzer through a step-by-step process that is set up to use the responses from one step's analysis to feed and build on the subsequent step. Also, just as the typical project is approved in phases, the business case flowchart (explained in detail in the subsequent section) is set up to be iterative. By making it iterative, there is the opportunity to incorporate any information known (and any best guesses) into the analysis flowchart. More importantly, because we request that a project start using the analysis flowchart as early as Opportunity Framing, it gets the core project team thinking about the various aspects of the project in terms of a modular option.

The nice thing about analyzing an alternative project execution philosophy, like modularization, is that it does not significantly alter the basic efforts and deliverables in the early phases (Opportunity Framing and Assessment). What it does offer is an alternative second option on project execution.

Consider how the modularization option opened up alternatives in the following project analysis.

PROJECT ANALYSIS

A company is considering a project along the US Gulf Coast site (USGC) to expand the current manufacture of a product. They have three potential locations they are considering. Since it will be a similar process to their other plants and this is only a high-level analysis, most have assumed the project execution would also be like the previous plants.

Obviously, one of the criteria is product delivery and a means of easy access to the market. However, at this early stage in conceptual analysis considering only a stick-built approach, there may not be any consideration of opportunities in terms of supply of construction materials to the project site during construction (after all, they have always been able to find some way to get the large process vessels into the previous sites). Without a specific effort to evaluate alternative forms of this project execution, such details would be pushed to later in the analysis. But with the use of a simple modularization option analysis, the three site options can be additionally considered in terms of the ability to bring large or heavy pre-assembled parts of the plant to the project site.

This forces project management in the early opportunity framing phase to be inclusive in terms of other implementation options. More importantly, because it has identified some of the key concerns with pre-assembly, the decision on the best project site will have to include the potential impact of adequate versus inadequate transportation access to that site. Because of the potential size of the industrial modules, specific considerations must be made early to avoid a module-limiting decision on the site location.

Along with the early access considerations identified above, other potential benefits will automatically tag along when discussing the modular option (that may add to the benefits of the pre-assembly off the site). For example:

- Will a large percentage of pre-assembly help alleviate any issues with local craft labor in terms of quantity and quality that might prove beneficial in terms of overall project cost and or schedule?

- Will the fact that weather may preclude continuous construction (e.g., during hurricane season) and an alternative can offer additional benefits in terms of schedule for offsite pre-assembly somewhere else in the US or the world?

• Is the physical configuration of the project process such that extensive work at elevations is required during construction (along with all the inefficiencies with respect to work productivity and safety) which could be reduced by prefabrication at a fab yard that has the equipment and experience to build at grade?

These should be subsequent issues and conversations that should immediately follow this line of alternative project execution thinking.

And, yes, this works both ways. Should, for some reason, the project analysis during these early discussions run into issues with the further development of large plug-and-play modules, the core management team can immediately shift to an early evaluation of other pre-assembly options. For example, parts of the complex process are very large in diameter and lightweight (e.g., a forced draft-fired heater). Typically, these are stick-built, but they can easily be adapted to varying forms of pre-assembly, depending on the type of heater and potential shipping envelope to the site. So, while modularizing the entire fired heater may not be an option, there are numerous ways to still pre-assembly parts offsite and ship them in, providing the project some benefits.

Again, such decisions are determined or "driven" by the project drivers mentioned earlier in this chapter. These must be identified and prioritized early in the project development. This business case flowchart helps surface and forces such evaluations in a proactive manner.

THE 13-STEP BUSINESS CASE FLOWCHART

The Business Case Flowchart (see Figure 5.4) consists of the following 13 steps:

1. Modularization Technical Feasibility

2. Identify Module Drivers

3. Analyze Module Potential

4. Perform Options Analysis

5. Develop Module Scope

6. Develop Module Size

7. Produce Module Definition and Index

8. Develop Execution Strategy and Execution Plan

9. Produce a Definitive Cost Estimate

10. Produce a Definitive Schedule

11. Check Module Viability

12. Proceed with Modularization to the Next Project Phase, or

13. Fall Back to Stick-Built

The following sections will further explain the 13 steps in detail. Keep this page referenced for ready access to the overview.

5.5 The 13-Step Business Case Flowchart

Step 1 Modularization Technical Feasibility

• Can you modularize or pre-fabricate and pre-assemble all or part of your project?

The answer should be a resounding "YES!" Even for the largest earthmoving or heavy excavation projects, there are always opportunities for some amount and type of pre-assembly.

Because we truly believe that there are opportunities for pre-assembly in every project, red flags should go up if the project core team says "NO" to any type of modularization or its analysis. It could mean there is a lack of understanding of the philosophy of the modular option and why it should be considered. It could mean there is an unwillingness in the project core team to move out of their current stick-built comfort zone of how projects are currently implemented. Following a status quo project pathway in any shape or form in this competitive industrial environment without evaluating alternatives, such as modularization, can be a harbinger of project disaster. It also may reflect a less than total commitment to an optimal project solution. In any case, a more detailed explanation of the value of modularization is required to move forward.

However, we left it in the business case model as a decision point primarily for the one-off project that genuinely might be simple or unique enough not to have any

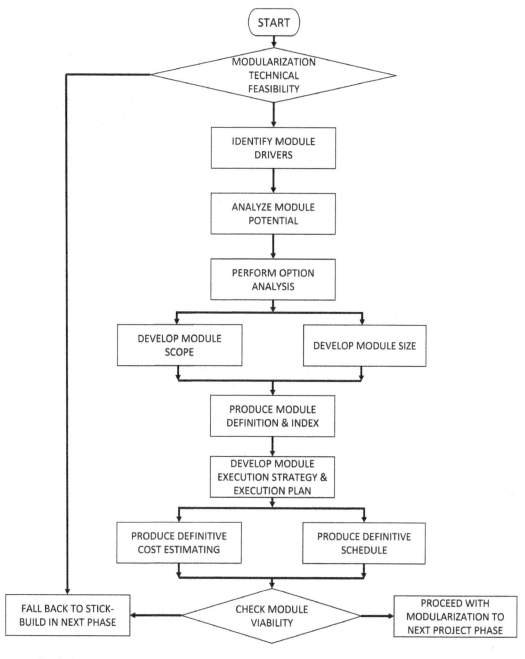

Figure 5.4 Business case flowchart.

great pre-assembly options. But it is understood and expected that such a decision is critically analyzed and not taken lightly.

Step 2 Identify Module Drivers

• Identify module drivers, rate and rank them, and then determine what solution best satisfies each of them.

An initial determination of precisely what is "driving" a project is crucial to identifying the best approach to project implementation. Earlier in this chapter, we examined several common and critical project drivers and explained why a modular or pre-assembly solution offers a better opportunity to satisfy that project driver. One thing that should be evident by this point in the analysis is that project drivers vary from job to job. There will always be a few that come up on every project (e.g., cost & schedule). So we tried to provide a basic understanding of these more common drivers in terms of how modularization will help meet these needs.

The important goal of all the previous discussion, explanations and examples is not how to argue both sides of the cost vs. schedule issue, but to understand that to finally get to the optimal project solution in terms of its construction execution, is to be able to identify ALL of the project drivers—the drivers known only to the owner/client and the ones known only to the EPC contractor. It is this complete picture that is required to make the best decisions.

From this flowchart standpoint, the discussions on project drivers should begin in Opportunity Framing as joint owner/client, EPC, and supplier discussions. All core team members should be willing to share what is driving the project from their personal standpoints. They should also be open to indicating which of their driver details are considered "off-limits" for the other project members. With regard to those, they should agree upfront to going back, internally analyzing these more confidential drivers, and committing to coming back to the group with results provided in a form that can be incorporated in terms of how they rate and rank with the other drivers. There should be enough transparency in how these results are supplied that the project team can take these driver analyses and seamlessly fold them into the rest, resulting in a complete project driver analysis.

Why such a push for completeness and transparency? The more "singleness of purpose" that a project can identify with, the more project energy will be channeled into that purpose. The more plugged in that the rank-and-file project members are, the more they can offer creative and innovative solutions. A project-wide understanding of what the project "goal" is and why it is being pursued a certain way will help ALL the project members understand how their piece of effort fits into the large picture. But to do this, the project needs a clear picture of what is driving it.

Step 3 Analyze Module Potential

- Complex analysis incorporating inputs from design/site/fab yard/cost and schedule.

Module potential could be considered the "yang" to the project drivers' "yin" (see Figure 5.5). Probably not the

Figure 5.5 Module potential and project drivers.

best analogy, but the symbol and the opposites it portrays fit nicely into how the two complement each other. The project drivers seemingly consist of all the project problems and challenges (negative and dark) of a project. The module potential consists of solutions and freedoms to these (positive and light).

The truth of the matter is that now that we have developed a comprehensive understanding of the project drivers (in step #2), we need an equally complete picture of the module potential so that the two can be juxtaposed for the best fit. This is the basis for the next step in the module flowchart. From an overall philosophical standpoint, as the business case is the heart of the modularization analysis, the module potential is the heart of the business case flowchart. To properly develop the module potential, the following four major inputs need to be added to Analyze Module Potential (see Figure 5.6).

Step 3.1 Design Inputs

Depending on whether the project is greenfield (new) or brownfield (existing), the design and layout will be directly impacted.

Step 3.1.1 New/Greenfield Site A greenfield site provides a clean sheet of paper in terms of layout and process configuration. However, there is always the tendency to make the layout exactly like or very similar to existing similar facilities with changes only due to location and type of inputs into the site and the corresponding product outputs.

Step 3.1.2 Existing/Brownfield Site A brownfield site is more challenging because, as part of the project premise, there is the goal of somehow fitting the square peg of the facility layout into the round hole of

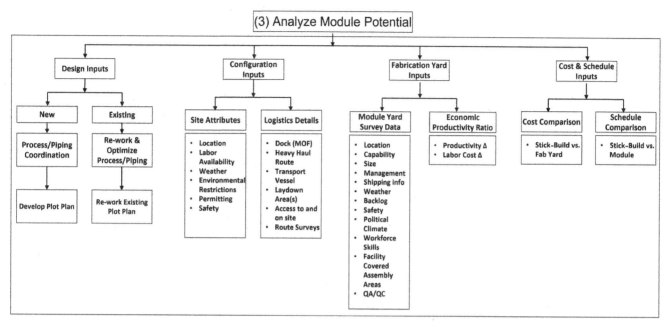

Figure 5.6 Four major inputs for module potential analysis.

the brownfield location. This goal or restriction requires that the process layout be critically evaluated in terms of how to get all the pieces into the proposed location, and so adjustments to the past or similar design layouts are understood to be required.

However, in either case, when performing the design development, do not fall into the trap of assuming that a stick-built layout is efficient enough to simply put "boxes" around areas and turn them into de facto modules. The stick-built layout is driven by different criteria than the equivalent modular layout, and boxing the stick-built configuration in many cases will result in an inefficient module configuration. The layout should be unique for several reasons, including the opportunity to use the third dimension (module height) to provide an additional degree of freedom or the opportunity to re-arrange the equipment. The result is typically a savings of space and, along with that, a saving of materials.

Step 3.2 Configuration Inputs

These inputs refer to the actual project site (site attributes) and access (logistics) to it.

Step 3.2.1 Site Attributes During the discussion on the universal key project drivers earlier, several of those are driven by the actual site location. Those site attributes include, but are not limited to:

- location
- labor availability
- weather
- environmental restrictions
- permitting
- safety.

The identified project drivers should be used as a guide in analyzing the various project site locations. Table 5.1 shows an example of module solutions for typical project drivers parameters.

Table 5.1 Module solutions for typical project drivers.

Project driver	Site parameter	Module solution
Cost	Limited workforce with good craft skills	Modularize—where labor is available
Schedule	Difficult and time-consuming site prep	Modules arrive later/more time for site prep
Logistics	Hard or expensive to get workers to the site	Use less expensive fab yard craft workers offsite via modularization
Environment	Pristine forests or fragile ecosystem	Modules reduce construction footprint at the site
Weather	Limited access part of the year	Build in temperate climate and ship to site

Step 3.2.2 Logistics The logistics details go hand in hand with the site attributes. Logistics are one of many site attributes, but since they encompass more than just the site and its vicinity, it is important that logistics be a separate consideration. These factors related to logistics include:

- Wharf or quay (materials offloading facility, MOF)

- Heavy haul route

- Transport vessel

- Laydown area(s)

- Access to and on the site

- Route surveys

For a stick-built project, the logistics typically deal with the determination of a pioneer camp and dock (to get the first of the materials and workers to the site) followed by a more permanent project materials offloading facility (MOF) or a road and river infrastructure stout enough to support the literal tons of bulks being shipped to site as well as the few large pieces of equipment that must be brought in by special carriers.

For the module job, all the above also pertain, but in addition, there must be considerations for the movement of much larger modules. This involves further analysis of the methods of first getting the largest of them to the vicinity of the site (by ocean-going transport or river barge), developing a much stouter MOF or wharf capable of receiving these in terms of weight and size, a heavy haul road with the load capacity and width to accommodate these large structures, and project equipment layout that will accommodate the movement of these large modules through the partially completed plant, and finally, a solid path of construction in terms of sequencing the delivery and setting of these large modules.

All this adds complexity to the analysis. It adds additional site constraints in terms of temporary access to and through the process units and a definite sequencing of construction that must be followed in order to be able to set all modules. This makes the site analysis a very early requirement in terms of even the initial site selection.

In short, any proposed site should have inherent access in terms of a proposed modular option as well as be able to internally accommodate a modular layout and delivery sequencing within the site boundaries.

Step 3.3 Fabrication Yard Inputs

The fabrication facility should match the project requirements in terms of project cost benefits as well as scope and capabilities. This is determined by a multi-faceted module yard survey with special emphasis on fabrication cost and productivity data.

Step 3.3.1 Module Yard Survey Data The data that need to be collected through the module yard survey include:

- location

- capability

- size

- management

- shipping information

- weather

- backlog

- safety

- political climate

- workforce skills

- facility covered assembly areas

- QA/QC

With fabrication facilities of all sizes and capabilities located in all parts of the world, the task of selecting the "best" in terms of a specific project can seem a bit daunting. Suppose you are a large EPC company with many years of experience. In that case, you probably can choose from a long list of previous projects that have worked with fabrication yards of various sizes over different parts of the world. However, this becomes more challenging if you are just starting out or your company does not have an offsite fabrication history. So, a couple of general guidelines in making a first cut on who or where to consider:

1. **Size matters.** There should be a good correlation between the total size of the module job (in terms of tons) and the capacity of the potential module yard.

Strive to select fabrication yards that are large enough to be able to handle your proposed capacity and not be overwhelmed by the required delivery timing. But make sure that the yard is not so big that the fabrication yard management does not ignore your "small" job. In other words, "not too big and not too small, but just right."

2. **Complexity matters.** Fabrication yards excel in different types and complexities of modules. Some are best suited because of their location and aerial extent for fabrication of the pipe rack modules that are relatively simple to assemble but require a large layout area for this assembly. On the other end of the fabrication spectrum, there are those highly sophisticated fab yards that have the capabilities to assemble the very complex modules and provide services in terms of equipment fabrication, procurement of bulks, detailed design, and plate girder and pipe spool fabrication. Often, these facilities are also located in areas where laydown space is a premium, so they are best suited for the fabrication of highly complex but compact large modules. The proposed size and complexity of a project's modules should be similar to what a fabrication yard typically produces. Overly complex modules may strain the technical competence of some yards. Conversely, overly simple and expansive modules may take up too much precious space in other highly sophisticated fab yards. In both cases, the project will suffer in terms of schedule delays or increased costs.

3. **Location matters.** If your project has a domestic US site, there is a tradeoff between the cost savings of fabricating overseas in the Middle East, SE Asia, or the Far East and the expense of the proper supervision or oversight at that fab yard in terms of the extra construction management team (CMT) plus the increased shipping costs from a fab yard halfway around the world. The module scope, along with the complexity mentioned above, will play a part in determining whether it makes more economic sense to spend a bit more in terms of module cost/ton and use a local yard that would minimize CMT as well as transportation costs. This means developing a good idea of the potential range of module scope options there may exist in terms of the project.

We will discuss this in more detail in Chapter 8.

Step 3.3.2 Economic Productivity Ratio (EPR) This step determines the true fabrication cost delta between work at the site and the same work accomplished at the module fabrication yard. This potential for cost savings drives projects to modularize and fuels a project's ability to spend more on a modular option in terms of engineering, materials, and shipping but end up with overall savings and benefits to the project. It is the simple concept of developing the relative work-hour cost for a certain amount of "production" in terms of a number that can be compared to the cost of other options that produce a similar amount of "production." The ratio of the two relative work-hour costs developed is called the economic productivity ratio (see Figure 5.7, for example).

Before we explain Figure 5.7, let us explain the definitions of the terms used:

Labor Productivity Factor: This is a frequently used but, unfortunately, greatly misused factor in the industry. And, to make it more complicated, depending on who you talk to, the increase in this number can reflect either improving productivity or declining productivity. So, without going into all the details needed to successfully explain our way out of developing an entire chapter on construction labor productivity factor and how

Description	Project Site	Current US Gulf Coast Site (USGC)	Selected Fab Yard
Labor Productivity Factor (Hours required to complete one unit of work; based on Ideal USGC = 1.0)	2.5	1.4	1.3
All-in Wage Rate (AIWR) (local cost / work-hour)	$135	$105	$45
Relative Work-hour Cost (RWC) (cost / hour for a similar unit of work)	$337.50	$147.00	$58.50
		Project Site to Current USGC	**Project Site to Fab Yard**
Economic Productivity Ratio		2.30	5.77

Figure 5.7 Economic productivity ratio example.

to calculate it, suffice it to say that some type of productivity measurement against a base case is required in order to be able to develop the EPR we are talking about.

In the case of the above example, we have assumed that the historical base productivity along the US Gulf Coast site (USGC) is 1.0. With the basis for this productivity in mind, we compared the anticipated productivity at the project site. We found that it takes almost 2.5 hours to produce or complete the same unit of "product" compared to 1 hour needed on the ideal USGC site. To the same extent, we found that it takes about 1.3 hours to produce or complete the same unit of "product" at the fab yard compared to the ideal USGC site.

All-In Wage Rate (AIWR): This is the total cost per work-hour (sometimes known as the "all-in" wage rate). One must be careful about how this work-hour cost is defined, but for this comparison, we suggest that the "all-in" wage rate be used, as it will be more reflective of the true total craft labor costs. The "all-in" wage rate includes not only the rate paid to the worker, but also all the other costs associated with supporting the worker on site—things like indirect labor, taxes, mobilization (mob) & demobilization (demob), local transportation, small tools, camp costs, contractor's field management, home office management, PPE, consumables, construction equipment, materials handling equipment, Quality Assurance & Quality Control (QA & QC), and contractor profit. As a side note, with all this added in, the "all-in" wage rate could be anywhere from 3 to 5 times the actual dollar amount the craft worker sees on their paycheck.

Relative Work-Hour Cost (RWC): This relative work-hour cost can be calculated by multiplying the labor productivity factor and the all-in wage rate (AIWR). This is a real cost rate that is required to complete a similar unit of work.

Economic Productivity Ratio (EPR): The EPR can be calculated by comparing the RWC of the project site to the RWC counterparts of the current USGC or the RWC of the fab yard. For example, in Figure 5.7, the EPR of the project site to the current USGC is 2.3 (= $337.50/$147), which can be calculated by dividing the RWC of the project site ($337.50) by the RWC of the current USGC ($147). In the same way, the EPR of the project site to fab yard is 5.77 (= $337.50/$58.50).

The EPR of the project site to fab yard is a number that shows when the project site and module yard relative work-hour costs (RWC) are compared, how many times more efficient the module yard labor is than the labor at the construction site considering the AIWR difference. The authors would like to state that this EPR concept is

adopted from the Construction Industry Institute (2012). Now, let's take a closer look at Figure 5.7. This example assumes that the project site is located in a cold or remote area, where lengthy nonproductive commutes exist from the accommodation worker camps to and from the site and productivity reductions occur due to the adverse weather conditions (such as extreme cold). The second row in Figure 5.7 shows that the labor productivity factor (hours required to complete one unit of work, based on the ideal USGC = 1.0) at the project site, the current USGC, and the selected fab yard are 2.5, 1.4, and 1.3, respectively. The productivity numbers mean, by just looking at the labor productivity factor (hours required to complete one unit of work compared to the ideal USGC labor productivity (1.0)), the project site requires 1.78 (= 2.5/1.4) times more work-hours than the current USGC site or the current USGC site will require only 56% (= 1.4/2.5) of work-hours compared to the project site. In the same way, we can find that the selected fab yard requires only 52% (= 1.3/2.5) of work-hours compared to the project site, or the fab yard is almost 1.92 (= 2.5/1.3) times more productive. On top of these labor productivity factor differences, there is the work-hour cost disparity between the "all-in" wage rate of the project-specific craft worker (at $135/work-hour) versus the typical all-in wage rate for some of the Far East fabrication yard craft (at $45/work-hour). As a result, this example case depicts a situation where the module fabrication yard is working almost twice as fast (2.5/1.3) with labor that is 3 times ($135/$45) as cheap.

Now, we need to see the effect of the combination of these two factors (labor productivity factor and AIWR) by looking at the Economic Productivity Ratio (EPR).

In this example (see Figure 5.7), as we already calculated, the EPR of the project site to fab yard is 5.77 (= $337.50/$58.50). This means the module yard is more than 5.77 times as economically productive as the project site. Putting it another way, every craft work-hour taken off the project site will support over 5.77 craft work-hours at the fabrication yard. As a result, the module fabrication yard will expend only 17% (= 1.0/5.77) of the estimated site cost (for an equivalent scope). With this large difference, it looks like the project should strive to maximize the reduction of work-hours at the site by moving them to the module fabrication yard.

What else does this example show? The EPR of the project site to the current USGC is 2.3 (= $337.50/$147), which means the current USGC is 2.3 times more productive than the proposed project construction site. Also, note that the module yard is more than 2.51 (= $147/$58.50) times more productive than even the current USGC construction site, which can be obtained by comparing the current USGC ($147) to the fab yard ($58.50).

Step 3.4 Cost & Schedule Inputs

These are typically worked together because, as previously mentioned, they are often discussed together. With these two project drivers as strong as they are, any modular analysis must develop some granularity on both of these as soon as possible.

Step 3.4.1 Cost Comparison In many cases, the initial cost estimates of a proposed plant are based on something historic—a previous plant or process or some other similar facility that is close enough to the desired project plant that it can be used. If the plant has older cost numbers, these can be updated and inflated to reflect present-day values.

With respect to assuming an estimated extent of modularization, a very early effort should be made to quantify this. Since many of the details of the equipment, layout, piping configuration, etc. are unknown at this point in time, as a first pass, assumptions can be made in terms of possible percentage total modularization. These should be based on how the project driver analysis performed in Step #2 reflected potential benefits from the use of modularization analysis and results to date identified from the analysis being performed in this step. For example:

1. If the project drivers indicate that a modularization option could provide significant benefits, then the assumption should be that the project will strive to maximize the percentage of modularization. From a practical standpoint, this could result in the potential removal of between 40% and 70% of the total stick-built scope (in terms of craft work-hours) from the site and sending it to a module fabrication yard.

2. If the drivers indicate no advantage (or very marginal advantage), we would suggest a quick review of the methods used in obtaining this conclusion as it could

mean that the project driver analysis was not comprehensive in terms of evaluation of ALL the potential benefits. But if it truly was the case that modularization did not provide an advantage, then the percentage modularization should be limited to no less than 10% of total stick-built work-hours. Such an example would be a project that limited the pre-assemblies to the obvious efforts, such as building the motor control centers (MCCs) and remote instrument buildings (RIBs) offsite as well as some or most of the pipe racks. It is suggested that this be the absolute minimum, as all of these are obvious choices and will almost always yield cost savings, just not great ones. Also, with this minimal approach, there would probably be little equipment optimization in terms of re-arrangements, and the plant would have a layout very similar to its stick-built predecessor.

3. If the drivers indicate some advantage (as is the case with most of the projects analyzed), then the initial assumption should be that the project would strive for the removal of between 10% and 40% of the total stick-built craft work-hour from the project site and this scope to be moved to some type of offsite pre-assembly. A typical example of such scope would include all the pre-assembly efforts mentioned above for the limited advantage option (MCCs, RIBs, pipe racks, typical skid-mounted equipment, vessel dressing) plus some equipment skids and a couple of major process modules. As the modules are defined for the modular option, efforts should be made to re-define the go-by stick-built plant in terms of optimizing the equipment layouts to get more sub-systems into potential modular groupings. This serves two great planning purposes:

    - It begins the thought process of how to optimize the sub-process location of equipment in case there is a decision to make these group modules (which can still provide benefits in terms of the stick-built layout if the decision is not to modularize).

    - It also begins the thought process of how the installation sequencing of these sub-process units might occur (which again can provide planning benefits whether you decide to modularize or stick-build).

The initial cost analysis for potential savings could be as simple as taking the anticipated percentage modularization times the estimated savings in scope cost between the cost of craft work-hours needed to do it onsite versus the cost of performing the same scope offsite in a module yard. In subsequent phases of the project, as more definition of the project process is developed, the cost analysis would move from this total percentage modularization to begin identifying the specific modules by size and shape, along with a finer-tuned estimating process that looks at the varied composition of structural steel to piping to equipment within each of these proposed modules. These would be compared to the cost of their equivalent stick-built portion of this plant.

Of course, since specific details on exactly what is in each module in terms of piping, electrical, and instrumentation will not yet have been developed in the early phases, this effort would be limited to something along the lines of identification of the major equipment in each module. Initially, the piping layout group (in conjunction with the process group) would take the equipment and lay it out in three dimensions to come up with a set of sized modules. The size of each would provide an idea of the total weight of the module based on typical volumetric analyses. If no other information was available in terms of other materials, this total module weight could be factored into how much was structural steel, piping, equipment, and to a lesser extent, the electrical, instrumentation, and controls associated with this equipment. Again, this factoring would be initially based on historical ratios of weights within the module—structural steel, equipment, and piping to the total module weight, structural steel to piping, piping to equipment, and other common ratios with all of these ratios varying depending on the type of module being considered (pipe rack vs. equipment, truckable vs. mega).

Once calculated, the cost again would be analyzed in terms of a delta cost savings between the proposed module scope and its equivalent stick-built portion of the plant. (More to come on this method in Chapter 9.) As the project progresses into the next phases, the detail on design becomes clearer through the development of block flow diagrams (BFDs) and process flow diagrams (PFDs). Historical information may be available in terms of similar parts of previous plants where the current design may not have progressed. At the end of the day, all these details on equipment type, number, and sizes are compiled into what is referred to as the "sized equipment list."

At this point in the project development, the plant process will have been defined to the level of detail that estimating programs such as the Aspen Capital Cost Estimator™ (ACCE) (aspentech, 2022) can be used to further refine the cost differences between stick-built and modular. Estimating tools like ACCE take the major equipment inputs from the sized equipment list and process information from the PFDs, and develop bulk quantities and costs for the equipment, the structural steel supporting the equipment, the process piping connecting the equipment, and the instrumentation, controls, and electrical required to start, operate, and control the equipment. What makes these tools particularly handy is that they develop imaginary volumes around the piece of equipment (e.g., vessel, pump, compressor) and include all the structural, piping, electrical, instrumentation, and even fireproofing within this volume. Then, by "stacking" and adding these equipment volumes, one can get an idea of the total volume of that particular module based on the proposed equipment to be incorporated into it.

Of course, this volume is only as good as the information fed into the program to develop it. But it provides the next level of detail over the volumetric determination described in the previous estimating phase. This can provide a reality check in terms of module size, which would adjust the cost estimate. And, with the additional cost and weight details on the steel, piping, and EI&C bulks provided by this estimating tool, potentially a more accurate description of the module can be developed.

Step 3.4.2 Schedule Comparison   The schedule is typically developed initially by reviewing the schedule of a similar plant and making initial adjustments to it for the modular case. As a starting point, if no other schedule information is available in terms of an alternative to stick-building, a good starting point for comparing the modular option is to assume a 10% or 2–4 months duration extension over the stick-built for the modular schedule.

There are a few reasons for assuming the first module schedule is a bit longer than the stick-built schedule.

1.   Many companies have found that this is the case—a module job just takes a bit longer than their best stick-built effort.

2.   If the scheduled delivery of the equipment and materials for the module job is similar to that of the stick-built

job and everything else is essentially the same, there are still the additional duration differences on the module job that will make it longer (e.g., the shipping of the completed module to the site and the time to set it in place and hook it up).

3. In some cases, the owner/client approval process does not allow pre-investment in engineering, which may be required to get an early start on the module structural design. As a result, the module fabrication start may be delayed. Since the module fabrication is often on the project critical path, this becomes a day-for-day schedule extension.

Of course, there are mitigating circumstances that can help offset each of these base assumptions. Still, as a first pass at a schedule comparison, the longer module schedule will force the more careful analysis of the entire suite of module benefits should the project be schedule-driven.

If this initial extra duration assumption for the module schedule is sufficient to kill the module option, then additional analysis and conversations with both the site construction team as well as the module fabricator should be initiated early to see what benefits from modularization have not been identified (e.g., a longer duration to work underground and foundations, module schedule surety, doubling up on module assembly and delivery, less duration for commissioning, more pre-commissioning at the fab yard, etc.). In addition, the owner/client needs to be brought into such a discussion to determine what can be done from their perspective to allow the EPC to start critical activities earlier (e.g., early approvals, pre-FID funding for critical path engineering design activities, etc.). This effort should result in a more module-friendly schedule.

Step 4 Perform Options Analysis

• Evaluate three or more module options.

With all the major details identified with respect to project drivers and module potential, some calculating (or analyzing) needs to be performed to see if there is a particular percentage of modularization or type of modularization that provides the optimal achievement of the project drivers.

Typically, with little or no module details, this option analysis is initially limited to evaluating the cost deltas for a low or medium or high percentage of the project scope to be modularized. As a first pass, assumptions can be made in terms of the impact on the project cost and schedule, assuming a 10–20% effort or a 35–50% effort or a 60–75% effort in terms of modularization to total project scope.

In addition, several different module fabrication yards should be selected for analysis, depending on the project module scope and location of the project site. Typically, a good number to begin with is 3, with all being from different parts of the world and a mix of large and smaller fabrication yards. This will provide support for a potential options mix from lots of large modules to just a few smaller ones and everything in between. Remember, the aim in this step is to look at the major options in terms of number, size, and complexity.

THE MODULE BUSINESS CASE ANALYSIS TOOL

If you or your company are starting the module business case analysis entirely from scratch, there are two options:

1. If your company is a CII member, download the modularization toolkit from here: https://www.construction-institute.org/resources/knowledgebase/knowledge-areas/modularization/topics/rt-283. This toolkit is only available to the CII members.

2. Others, including students and academics, can download one of the authors' papers (Choi et al., 2019) entitled, "Modularization Business Case Analysis Model for Industrial Projects," which is available here: https://doi.org/10.1061/(ASCE)ME.1943-5479.0000683

The modularization toolkit was developed by MCBA (https://www.construction-institute.org/groups/communities-for-business-advancement/modularization), where both authors are leaders of the group and the lead tool developers. The first tool in the modularization toolkit is the modularization business case analysis tool. One of the authors enhanced the tool and published an academic journal paper (Choi *et al.*, 2019) in a renowned journal in the area of management in engineering. Thus, reading (Choi *et al.*, 2019) will be sufficient to understand the tool. The following explains the tool briefly.

The modularization business case analysis tool is an Excel spreadsheet-based cost comparison tool to be used as a first pass at this option analysis. While not as sophisticated as many private EPC companies' estimating tools, this tool was set up for first-time users and had the following attributes (Choi *et al.*, 2019):

- It is focused on identifying the optimal level of craft work-hours to move offsite.

- It is based on (Construction Industry Institute, 2012; O'Connor, O'Brien, and Choi, 2013; Choi and O'Connor, 2015).

- It prompts the addition of data for major project drivers.

- It accepts additional details.

- It is iterative and recommended to be used in subsequent project phase analyses.

- It is flexible and can be adjusted to meet specific project or client requirements.

- It provides specific savings, not just indicative values.

- It provides an idea of the expected values, which can be derived by going deeper into the details, where the actual values can be calculated.

The tool prompts input based on the business case flowchart, requesting key data requirements (Choi and O'Connor, 2015; Choi *et al.*, 2019), such as:

- project total installed cost (TIC)

- project schedule savings

- percentage of modularization

- relative work-hour costs

- labor productivity

- modularization drivers and benefits.

Most EPC companies have some type of cost analysis program that can be modified slightly to capture these delta costs. If not, there is commercial software, including the Aspen ACCE, which has a cost-estimating section specific to a module project.

The result of this analysis is a graphical representation of the potential savings that can be realized by that particular set of cost and productivity options, along with the summation of the values of all the soft issue benefits of going modular. The tool inputs are shown in Figure 5.8,

and the outputs are shown in Figure 5.9 and Table 5.2 to give the readers an idea of the simplicity of the analysis.

The key to this analysis is to be able to place dollar values not only on the tangible differences, such as cheaper but more productive labor, additional engineering requirements, additional shipping costs, installation at site, but to be able to also identify a dollar value for all of the soft benefits, such as safety, quality, craft skills, working at grade, working in a temperate climate, schedule surety, etc. While it may not be politically correct to place a dollar value on safety or community relations, there is one, and there should be an effort to identify it because the sum of all the hard and soft benefits equals the actual cost delta of taking the module alternative. If you are interested in learning more about the tool in detail, please refer to (Choi *et al.*, 2019).

Step 5 Develop Module Scope, and Step 6 Develop Module Size

• Identify the best option determined in Step #4 in terms of a definitive scope.

Step 5.1 Module Scope

This is where the early percentage option that provides the optimal project solution actually must be turned into a definitive reality. Typically, this takes the form of a plot plan or some other general layout where the module scope can be identified in terms of area and extent.

So, why are these two steps considered separate from each other and separate from the option analysis? This is necessary because running a calculation that shows the optimal cost benefit is, for example, 75% modularization and identifying all the equipment, labor, and materials required to do so and placing them into a module configuration that will achieve this 75% modularization are two completely different activities. The challenge, especially when working to achieve the higher percentages of modularization, is finding the concentrated pockets of construction craft work-hour requirements for completing an area and then bringing these together in proximity to each other to permit them all to be incorporated into a module.

So, how is this done? During the early project phases, the project management must rely on the historical information regarding craft work-hour requirements to complete different areas of the project plant for this analysis. Then, working with the process experts (who understand what can be moved around with respect to the process flows) and the piping experts (who can lay out some

alternatives in terms of connecting these areas together), one can start the evaluation of what arrangements in the sub-processes can be laid out that will be able to concentrate the craft requirements. This is critical since the aim is to transfer craft work-hours from the project site to an offsite area like the module fabrication yard.

The pictorial plot plan development example (Figures 5.10 and 5.11) shows how the equipment within a sub-system can be located "all over the map" (Figure 5.10) and how the concentration of craft work-hours (site labor intensity) also varies from one group of equipment to another (Figure 5.11). Figures 5.10, 5.11, and 5.12 are the same as Figures 4.9, 4.10, and 4.11. The exercise required is to find as many of the high labor intensity pieces of equipment and areas as possible that can be gathered in the same proximity to help define a potential module.

By considering BOTH the site craft labor work-hour concentrations and the sub-process equipment groupings when combining into modules, the ideal result is sub-process-specific modules that are able to incorporate the maximum site work-hours possible. The result in this stylized example where these two goals are worked together is the arrangement of the six sub-systems into a geographically centralized area that was consolidated and realigned based on the craft labor density data into the final "six-pack" module configuration (see Figure 5.12). As an added benefit, it turns out that when you can consolidate and centralize the sub-processes and "stack" them on more than one level, the actual footprint of the plant is decreased, sometimes substantially, as noted in Figure 5.12, where the final module configuration and sizing have been overlaid on the original stick-built plot area of Figures 5.10 and 5.11. This comes in handy where the plant site space is a premium.

Step 5.2 Module Size

For an unrestricted module shipping envelope, this is the direct result of the work done in concentrating the craft labor requirements and sub-processes into areas. However, a project seldom has an unrestricted shipping envelope. So,

Modularization Business Case Analysis Tool - Input1			
Instructions: If you are using the tool for the first time, it is recommended to start inserting the data from the top to bottom. If you are familiar with modularization business case analysis, it is recommended to start from right (the Third level questions) to left. Bolded boxes are recommended (minimum) data for the optimum analysis output. Other normal boxes will be calculated if the bolded boxes are completed. The user may decide to complete data on normal boxes as well if he/she wishes.			
Business Case Analysis Steps	First Level Questions		
0. Project Info	0.1 Project Name		Project XXX
	0.2 Estimated Stick-built TIC	$	200,000,000.00
	0.3 Target % Modular	%	25%
1. Technically Feasibility Analys	1.1 Technically Feasible?	Y/N	Y
2. Schedule Benefits Analysis	2.1 Expected Schedule Saving by MOD	$	400,000.00
3. Site Survey	3.1 Relative MH Cost at Site / MH	$	275.00
4. Module Yard Survey	4.1 Relative MH Cost at Assembly Yard / MH	$	66.00
	4.2 Relative MH Cost at Fab Yard / MH	$	99.00
*Cost Benefit Analysis	Baseline Cost - Module Portion	$	50,000,000.00
	Module Cost	$	45,681,250.00

Figure 5.8 Modularization business case analysis tool example (input page).

while trying to group equipment into modules, one needs to remember the first criterion defined on a potential module project was the size of the shipping envelope, or put another way, how "big a box can be shipped to site."

So, in the above-described effort of defining module scope, in the back of everyone's mind should have been how big one can make the modules. Because while a smaller shipping envelope does not necessarily kill a module project, it does make it more challenging because instead of the potential six large modules from the example above, one must find a way to efficiently further subdivide each of these six sub-processes into several more modules to fit the shipping envelope without creating a

Instructions: Provided numbers are only for example. Several notes are left for the users information. The user can insert more factors and link to higher level.

Second Level Questions			Third Level Questions			
2.1.1 Expected Schedule Saving		40 Days				
2.1.2 Cost per day of Schedule	$	10,000.00 /Day				
3.1.1 Labor Productivity at Site		2.2				
3.1.2 Marginal Cost of Site Construction / MH	$	125.00 /MH				
4.1.1 Labor Productivity at Module Assembly Yard		1.2				
4.1.2 Marginal Cost of Module Assembly / MH	$	55.00 /MH				
4.2.1 Labor Productivity at Fabrication Yard		1.5				
4.2.2 Marginal Cost of Fabrication / MH	$	66.00 /MH				
Additional Installation Cost at Site	$	41,250.00	Required Installation Labor (MH)		150 MH	
Additional Shipping Cost	$	600,000.00	Module Transportation Cost	$	300,000.00	
			Pre-fab Transportation Cost	$	200,000.00	
			Transportation Study Cost	$	100,000.00	
Additional Fabrication Cost at Fab Yard (Mtl. & Labor)	$	13,090,000.00	Structural Steel Quantity Removed from Site		1000 Tons	
			Cost of Structural Steel	$	2000 /Ton	
			% Structural Steel Increase for Mod	%	10%	
			Required Structural Fabrication Labor (MH) / Ton		100 MH/Ton	
Additional Module Assembly Cost at Assembly Yard	$	26,400,000.00	Total Module Weight		4000 Tons	
			Required Module Assembly Labor (MH) / Ton		100 MH/Ton	
Additional Engineering Cost	$	5,000,000.00	Additional Eng. MH's for Mod		50000 MH	
			Engineering costs	$	100 /MH	
Other Cost for Module	$	550,000.00	Yard Management Cost	$	150,000.00	
			TAX	$	100,000.00	
			Import Duties Cost	$	100,000.00	
			Tran. Insurance Cost	$	100,000.00	
			Early Investment Cost (Finance)	$	100,000.00	
			Etc.	$	100,000.00	
5.1.1 Safety Benefit	$	100,000.00	On-site Historical RIR or SIIR			
			Off-site historical RIR or SIIR			
			Projected RIR or SIIR			
			Safe Behavior Culture			
			Hazard Awareness Culture			
5.1.2 Quality Benefit	$	100,000.00	Non-conformance Reports			
			Cost of Rework	$		
5.1.3 Benefits to Local Community	$	10,000.00				
5.1.4 Contingency Benefit	$	10,000.00				
5.1.5 Site Indirect Costs Benefit	$	100,000.00	Site Preparation	$		
			Site Temporary Facilities	$		
			Site Construction Equipment	$		
			Site Environmental Impacts	$		
			Site Restoration	$		
			Indirect Staff and Supervision	$		
5.1.6 Risk Benefit	$	100,000.00	Permit Issues	$		
			Environmental Impact	$		
			Weather Impacts	$		
			Political	$		
5.1.5 Other Expected Benefits	$	100,000.00				

Figure 5.8 (Continued)

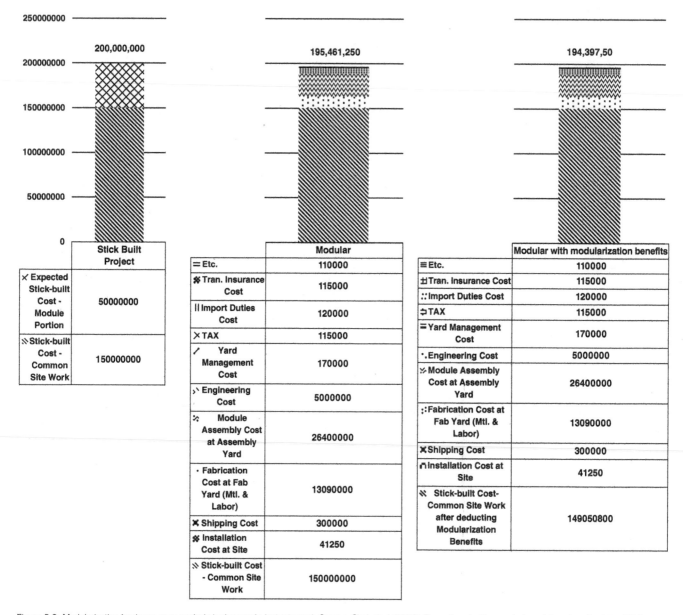

Figure 5.9 Modularization business case analysis tool example (output page). Source: Choi *et al.* (2019). Reproduced with permission of American Society of Civil Engineers (ASCE).

large additional delta craft workforce requirement at the site to re-connect the module after it arrives.

To summarize, module size is important. Where larger equipment and complex systems are present, larger modules are preferred over smaller ones in order to get complete sub-systems into a single module, allowing a higher completion percentage in the fab yard and reducing interface connections at the site. When the sub-systems allow for potentially smaller options, the authors favor these, assuming an equivalent number of site craft hours can be transferred to the fab yard. Smaller modules are faster to assemble, easier to ship, and offer more flexibility in terms of movement to site and setting at the site. If properly laid out, the additional interface connections at the site can be minimized.

Table 5.2 Modularization business case analysis results example.

Summary of cost results		
Results	Unit	July 10, 2017
Estimated stick-built TIC	$	200,000,000.00
Target % modularization	%	25
Common site work	%	75
Technically feasible	Y/N	Y
Relative WH cost at site/WH	$/WH	275
Relative WH cost at assembly yard/WH	$/WH	66
Relative WH cost at fabrication yard/WH	$/WH	99
Stick-built cost—common site work	$	150,000,000.00
Expected stick-built cost—module portion	$	50,000,000.00
Module cost	$	45,461,250.00
• Installation cost at site	$	41,250.00
• Shipping cost	$	300,000.00
• Fabrication cost at fabrication yard	$	13,090,000.00
• (materials and labor)		
• Module assembly cost at assembly yard	$	26,400,000.00
• Engineering cost	$	5,000,000.00
• Yard management cost	$	170,000.00
• Tax	$	115,000.00
• Import duties cost	$	120,000.00
• Transportation insurance cost	$	115,000.00
• Early investment cost (finance)	$	110,000.00
• Miscellaneous	$	110,000.00
Modularization benefits	$	1,849,200.00
Schedule saving benefit	$	900,000.00
Modularization benefits	$	949,200.00
• Safety benefit	$	150,000.00
• Quality benefit	$	150,000.00
• Benefits to local community	$	10,000.00
• Contingency benefit	$	20,000.00
• Site indirect benefit	$	150,000.00
• Risk benefit	$	369,200.00
• Other expected benefits	$	100,000.00
Estimated stick-built TIC	$	200,000,000.00
Estimated modular project TIC	$	193,612,050.00
Estimated total cost saving	$	6,387,950.00

Source: Choi *et al*. (2019). Reproduced with permission of American Society of Civil Engineers (ASCE).

Step 7 Produce Module Definition and Index

• Develop the actual list of modules.

Once the module scope and module size have been identified, this effort becomes a high grading of the initial pass made at trying to fit everything in the module. As it turns out, there are some pieces of equipment that are very well suited for modularization, and there are others that need to be left out of the module. In this step, these details are evaluated, and decisions should be made with respect to the final module layouts.

Special considerations should be given to any piece of equipment that is extremely large, extremely heavy, or has rotating parts. Any of these can create inefficiencies in design with respect to the module. For example:

• **Tall towers.** These create large moments at their base due to wind and seismic loading. Massive concrete foundations typically resist these moments. But in the case of a module, these same moments must now be resisted by the module steel, requiring excessively large steel girders or massive trussing to withstand these loads. Often, if the connections and amount of "dressing" or finishing of these towers are not excessive, it is more efficient to plan to set these towers just outside the module limits using the more conventional concrete foundation for support and resistance to these forces.

• **Heavy vessels.** While not as critical as the tall towers since they will not experience the large moment forces from seismic and wind, they still require massive steel plate girders to support and bridge the weight over to the main module column foundations. Again, this is not a showstopper, but it may be worth a second look at how critical it is to have this large vessel in the module versus just outside the module limits. This decision is again driven by how much more hook-up is required at the site to connect the vessel back into the module equipment and piping.

• **Rotating equipment.** This is the most difficult of the three to analyze. While small pumps and other rotating machinery pose absolutely no issues on a module, as the piece of machinery gets larger and the rotating forces increase, at some point, there will need to be an analysis of the interaction of the rotating equipment

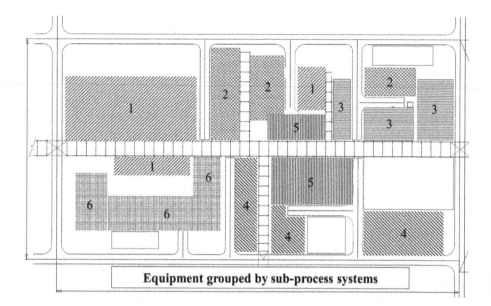

Equipment grouped by sub-process systems

Figure 5.10 Stylized stick-built plot plan outlining the six sub-process areas.

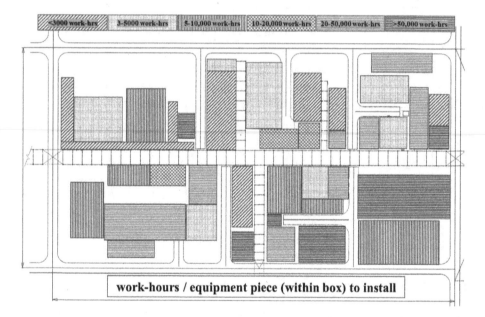

work-hours / equipment piece (within box) to install

Figure 5.11 Same project plot plan but showing stick-built site craft work-hour concentrations.

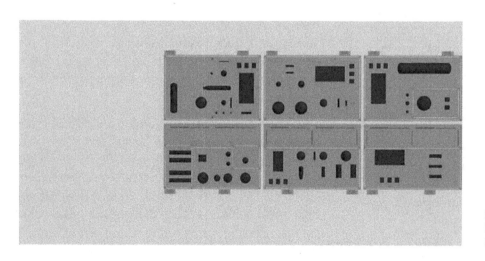

Figure 5.12 Equivalent module "six-pack" layout with each sub-process as a separate module.

with the structure in terms of vibration and damaging harmonics. Historically, very large compressors and other rotating machinery were set on massive foundations attached with special plates or rails that were epoxy-grouted into place. The epoxy grout was strong enough as well as "flexible" enough to withstand the vibrations and maintain a crack-free connection between the piece of machinery and its concrete foundation. In some cases, where these large rotating machines or compressors needed to be elevated to allow piping to be connected below the machine, so these large concrete foundations were elevated on massive legs, forming a huge "tabletop." The forming and pouring of these massive tabletop concrete structures comprise a very craft labor-intensive operation, so there has always been an interest in options that will provide an alternative to this site-intensive activity.

The module offers such an alternative; however, the dynamics of the machinery rotational forces must be accurately determined so that destructive vibration and harmonics do not develop. Besides the annoying vibration that could be transmitted throughout the module structure, there is the real concern of structural fatigue and premature fatigue of many of the minor connections on instruments and the wear and tear on the more delicate instrumentation itself. Historically, this analysis has been difficult and a bit of a black box solution with no guaranteed satisfactory results. As a result, many clients and EPCs fall back on the tried-and-true use of large concrete masses to withstand these forces. However, with the higher speed aero-derivative (jet engine compressors and turbines) type drivers, the forces to be designed for are less, and the industry trend is toward the installation of larger and larger capacity rotating equipment on the modules. So, while we expect to be able to incorporate larger and larger machinery into a module, this is still something that must be evaluated early and thoroughly.

As all these considerations are identified and analyzed, the module definition and index (listing) continue to be developed and refined. While it is not expected that the first pass at this effort will result in the final and complete listing, it is expected that this must be completed as soon as possible because this index forms the basis for many different project initiatives involving these modules, such as fab yard analysis, shipping considerations, HH road,

and logistics to the plant site, size and capacity of the MOF, and temporary module storage, to name but a few.

Step 8 Develop Execution Strategy and Execution Plan

• Draft the project execution plan incorporating modular concepts.

By the time one gets to this step in the 13-step business case flowchart, enough module-specific information has been identified to be incorporated into the project execution, which in many cases is still based on a stick-built strategy. Because of this need, we provide a definitive step in the flowchart specifically for this incorporation of module-specific information and the dedication of the time to critically identify the adjustments required to the project execution now that modules have become part of the process. With at least 107 execution plan differences (EPDs) identified by the CII RT-283 research (which will be further defined in Chapter 6), the development of the module execution strategy is no small task.

Initially, the project execution strategy should be critically reviewed in terms of the demands that the module effort has on the rest of the project continuity. For example, we know from the previous flowchart steps that we need to accommodate the modules in terms of earlier needs of the following:

• **Size**, as it pertains to early shipping and logistics requirements.

• **Engineering**, as it pertains to the early module structure and equipment layout.

• **Materials**, with respect to early bulk purchases, eventually required for initial module fabrication.

• **Equipment design**, where specific details are required to design supports and connection details on the structural steel.

• **Fab yard analysis**, where early identification of the proper fab yard is required for interaction on the design as well as initiating early contractual efforts required to develop a proper agreement.

• **Logistics**, where early efforts are required for the HH route right of way (ROW) as well as early contracting for the vessel shipments.

In the initial phases of the project, this execution strategy needs to have at least reviewed all 107 applicable "modular" execution aspects for incorporation into this strategy. As we will explain in more detail in Chapter 6, these EPDs cover 21 major topics spanning several project phases beginning in the opportunity framing phase and extending into the EPC phase. These 21 topics impacted by the module execution philosophy run the gamut from identifying project objectives to final system testing, commissioning, and start-up and are listed below to help frame the extent of the impacts (Construction Industry Institute, 2012; O'Connor, O'Brien, and Choi, 2013, 2016):

- Project Objectives

- Organization & Staffing

- Modularization Scoping, Layout Process, & Plot Plan

- Transport Route Study & Planning

- Planning & Cost Estimating

- Craft Labor Relations

- Contract Strategy

- HSSE & Social Impacts

- Risk Management

- Modularization Business Case Validation/Refinement

- Stakeholder Alignment and Reframing

- Fabricators, Contractors, & Subcontractors

- Procurement Strategy and Owner-Furnished Equipment

- Basic Design Standards, Models, & Deliverables

- Methods, Heavy Lifts, & Construction Facilities

- Procurement, Vendor Data, & Expediting

- Scope Freeze & Change Management

- Project Controls & Site Management

- Quality Assurance/Quality Control

- Detailed Design Deliverables

- System Testing, Commissioning, & Start-up

As mentioned, this incorporation is no small task. Its development should begin as early as the first pass at this 13-step process during the opportunity framing phase and continue until all details have been identified and resolved. While there was an attempt to arrange the list above in terms of "when" these items are initially identified and worked, there is no set or correct sequence for addressing or completing them. However, it is important that as soon as one of the items comes up, it be addressed in whatever level of detail currently available, if for no other reason than to create a placeholder or "tickler file" for completing when more information is available.

Step 9 Produce a Definitive Cost Estimate

- Ensure the project cost estimate incorporates all things modular.

This is the point in the 13-step flowchart where the "official" project cost estimate should be evaluated in terms of modular impacts. Yes, we performed a cost exercise in Step #3, but that effort was limited in scope and included as one of many areas reviewed at a high level to make an initial determination of the module potential. Step #3 analysis was not meant to be the all-inclusive project cost estimate. It was meant to be a "quick and dirty" analysis, emphasizing the main contributors to the cost benefits of modularization with the limited information available.

So, somewhere in this 13-step business case flowchart, there needed to be a deeper dive into the project costs with respect to modularization. Because, along with all the obvious benefits of modularization, there are potential downsides in terms of the cost impacts. Items such as the following are not incorporated in the Step #3 high-level cost study:

- Impact of early engineering costs for module design support (including risk or reward for these early expenditures).

- Impact of additional project risk in adopting a module philosophy (impact from late delivery due to module execution plan).

- Additional costs associated with the module fab yard oversight (construction management).

So, the cost estimate associated with this step is to further detail and complete the project cost estimate. In this step, the module team members should be plugged into the project cost estimate development to ensure that not only the obvious benefits and differences are included, but also the not so obvious ones. The differences between the cost estimate in Step #3 versus Step #9 are summarized below:

- The cost estimate in Step #3 is an internal analysis strictly limited to identifying the module potential.

- The cost estimate of Step #9 is the project cost estimate, an emphasis on the module impacts and their interaction with the project estimate to ensure the project estimating team includes everything modular into it.

Step 10 Produce a Definitive Schedule

- Ensure the project schedule incorporates all things modular.

Like the cost discussion in Step #9, the scheduling effort has the same approach and goals. The early scheduling efforts initially reviewed in Step #3 were strictly associated with how the change from a stick-built to a module execution philosophy would impact the overall project schedule. The details were not available and not fully addressed.

But with the project's adoption of the modular philosophy, the project team, with the input from the dedicated module team, needs to concentrate on the schedule to ensure that all things modular are included. With all the interrelationships between and among the various groups within the project, there needs to be a concerted effort at determining predecessors and successors in terms of the critical path schedule for developing a module ready design with purchase-ready materials that will meet the fabrication yard requirements. This means that the schedule development and review must be coordinated with leads that understand not only their needs but also how their deliverables impact the other groups.

Again, as with the cost step, the schedule effort in this step is more coordination of the schedule to ensure it truly is representative of the project schedule with a modular approach.

Step 11 Check Module Viability

- End of flowchart check on whether the module option is still viable.

There is not much to say about this step. But while it looks like a simple YES/NO toggle on the flowchart, it is more than that. It is a formal step set up to pause the business case analysis, re-evaluate all the work performed, and contemplate where this business case and the project are going. The reason for this step in the analysis is to determine if something is not working, either for a segment of the modularization scope or the entire project. This is when you officially say "stop"—this plan is not working for one reason or the other, and the project team needs to evaluate the concern officially and act on it.

Step 12 Proceed with Modularization to the Next Project Phase

- Team consensus to move with the current modularization option into the next project phase.

This is the affirmation step where the project team takes the currently identified module scope for this project, confirms it is the optimal module scope and moves it into the next project phase of detailed development. It is understood that there may be some adjustments to this scope in the next project phases, but everyone agrees that it is currently the best option at this time.

Step 13 Fall Back to Stick-Built

- Team agreement that part or all of the modular effort needs to be adjusted or reverted back to the more traditional stick-built approach.

This is the alternative to moving forward with the proposed optimal module solution. It is, in effect, saying that the project execution philosophy will be modified for the next project phase from what was considered the optimal modular solution to something else—typically less than what was currently being recommended.

So, how and why would this flowchart analysis need to revisit the conclusion identifying the optimal module solution for this project? It could be any one of a multitude of reasons, for example:

- The owner/client wants to accelerate the completion by building as the design is developed.

- The licensor is still developing the process and is expecting to continually make changes or adjustments to the process, causing major re-work throughout the execution phase.

- The site must be moved to another location with a more restrictive shipping envelope, precluding the previously optimized modules' size, number, or weight.

Of course, a move to step #13 does not preclude modularization in its entirety. The cause or causes that drove the project to say "no" to the optimal module solution do not eliminate the opportunity to suggest a smaller scope for pre-assembly. This step only opens the door to all the other alternatives to the current proposal. It is hoped that the reasons for deciding against the optimal module solution identified by step #12 are critically reviewed to ensure they are legitimate and no workarounds exist. Then, it is suggested that all the alternatives to the optimal module solution previously abandoned are equally thoroughly and critically reviewed in terms of their benefits to the project.

5.6 How Often Should the Business Case Flowchart Be Utilized?

Ideally, it should be initially used as soon as a project becomes an idea in the owner/client CEO's head. While many details may not be available, the identified ones need to be vetted against a modular execution philosophy. So, it should be started in opportunity framing and then reviewed at least once during each of the subsequent project phases: assessment/selection/basic design and potentially taken even into EPC. Additionally, review updates should be made when something significant occurs in the project execution planning that suggests a change in the project deliverables.

5.7 Summary

It is hoped that by this time, you have reached the conclusion that some amount of modularization is not only possible but beneficial to every project. With the module potential identified, the next step is to determine what the optimal module solution might be. The 13-step Business Case provided here is the road map for this identification. Its step-by-step approach provided not only a comprehensive guide identifying the major factors to be considered in this search for the ideal module scope

but also the suggested order in which to consider them. Then, for additional support, links to the Modularization Toolkit (including the Business Case Analysis Tool) developed by the CII Modularization Community of Business Development (MCBA), as well as links to an advanced version of the tool (developed by one of the authors), have been provided. In either version, this Excel spreadsheet-based cost comparison tool is suitable as a first pass at module option analysis. While not as sophisticated as many private EPC companies' estimating tools, it is a good initial set-up for first-time users and easily adaptable to any particular project specifics.

References

aspentech (2022) Aspen Capital Cost Estimator™. Available at: https://www.aspentech.com/en/products/engineering/aspen-capital-cost-estimator (Accessed: 3 January 2022).

Choi, J.O. and O'Connor, J.T. (2015) Modularization Business Case Analysis: Learning from Industry Practices Tool, In Al-Hussein, M. et al. (Eds.) 2015 Modular and Offsite Construction (MOC) Summit. Edmonton, Alberta, Canada: University of Alberta, pp. 69–76. doi:10.29173/mocs178.

Choi, J.O. et al. (2019) Modularization Business Case Analysis Model for Industrial Projects. Journal of Management in Engineering, 35(3), 04019004. doi: 10.1061/(ASCE)ME.1943-5479.0000683.

Construction Industry Institute (1986) Control of Construction Project Scope. Austin, TX: The University of Texas at Austin: Construction Industry Institute.

Construction Industry Institute (2012) Industrial Modularization: How to Optimize; How to Maximize. Austin, TX: The University of Texas at Austin: Construction Industry Institute.

O'Brien, W.J., O'Connor, J.T., and Choi, J.O. (2015) Modularization Business Case: Process Flowchart and Major Considerations. In Al-Hussein, M. et al. (Eds.) 2015 Modular and Offsite Construction (MOC) Summit. Edmonton, Canada: University of Alberta, pp. 60–67. doi:10.29173/mocs178.

O'Connor, J.T., O'Brien, W.J., and Choi, J.O. (2013) Industrial Modularization: How to Optimize; How to Maximize. Austin, TX: The University of Texas at Austin: Construction Industry Institute.

O'Connor, J.T., O'Brien, W.J., and Choi, J.O. (2016) Industrial Project Execution Planning: Modularization versus Stick-Built. Practice Periodical on Structural Design and Construction, 21(1), 04015014. doi: 10.1061/(ASCE)SC.1943-5576.0000270.

chapter 6 The Module Team and Execution Plan Differences

So far, we have talked about what modularization is, what modules look like, and, in Chapter 5, how to determine the best mix of modules for a project. That is a lot of new information and additional work that is not typical in a stick-built project, so the two essential questions are: "Who is accountable for all these newly acquired project responsibilities, activities, and goals?" And then, "What is the difference between the stick-built project execution and the modular project execution philosophies?" In this chapter, we will begin working on answering these two essential questions in terms of the module team and the execution plan differences.

6.1 The Module Team

Since the modularization effort impacts all aspects of a project, typically, there is really no one team member, group of people or single engineering discipline in the stick-built project execution philosophy that can be pointed to and assigned such an activity with the simple directive, "You! Handle it!" (Well, a project manager could delegate like this, but it might not work very well, even if accepted.) There is just too much background knowledge to develop and too many interrelationships to understand and manage to simply assign the task to a new person or group naïve in the modular philosophy.

The best approach to managing a modular project is to set up a module team. The size can vary from a single person (for the very small local project) to a multidisciplinary team of individuals (for the large international endeavor). Nevertheless, regardless of project size,

several important common requirements of this module team need to be identified and incorporated.

6.1.1 A Module-Savvy Leader

The module team leader must have had past experience in coordinating one or more module projects. At the very minimum, they must have at least been associated in a management role of some sort on a module project and worked on that project from beginning to end (inception to start-up). Why? We have previously mentioned (and will go into more detail later in this chapter) the over 100 execution plan differences between the stick-built and modular projects. Without the benefit of working on a module project from beginning to end, the proposed module lead would not have seen and would not appreciate all the nuances in terms of project deliverables, their interactions, and timing, and how they must shift from a stick-built project basis in order to meet the needs of the module fabrication yard initially, and ultimately the site construction team and startup and commissioning.

Reading about it (even in this book) is not the same as living through it (on a project). There is nothing like working on the daily challenges of coordinating information, materials, contracts, etc., on a day-to-day basis. (And only realizing later, using your much more accurate 20-20 hindsight, that you almost "screwed" it up.) But this first (or previous) module opportunity will have provided you insight into how you might handle things the next time. Speaking from experience, the second (and subsequent) module jobs went much smoother than the first because of these "learning" moments.

Of course, it does the project no good if this module-savvy leader cannot effectively communicate and truly lead the project into action based on project timing and sequencing they are unfamiliar with. So, this module-savvy leader must also command some people skills, be good at getting the module team together, and ultimately get the project moving in the right direction at the correct time.

6.1.2 The Module Team Members

Unless the module lead is a superhero and can work continuously and tirelessly, they will need a team. The team should be made up of "like-minded" individuals who either share the understanding of the modularization concepts and timing or at least are open to learning about them. In addition, because this project philosophy impacts every team and workgroup within the project, the team should have a representative from each of these workgroups actively involved. It requires a special person to be a team member—one willing to spend a bit of "extra" time above and beyond their "regular" job assignment to understand the modular philosophy, how it impacts the overall project, and in particular, how it impacts their specific workgroup. As the representative for that workgroup in the module team, each team member, who is the link from their group to the module team and the interface within their discipline or workgroup, to first identify changes that will need to be made and then (almost like a "double agent") report back to the module team on progress or issues. While not as covert as characterized by the term, it is important that feedback both comes from and goes to each workgroup on module interactions in a much more collaborative nature than what most project members are familiar with. So, each member of the module team ALSO needs to be a good communicator and a people person.

6.1.3 Critical Team Members

While everyone on the team has a part to play with respect to industrial modularization projects, the following members representing the following fields are critical to its success. They, therefore, need a good understanding of their particular roles.

6.1.3.1 Civil/Structural

Since this discipline essentially "leads" the engineering disciplines in terms of design requirements and early timing, a member who will represent them must have strong leadership skills and a good understanding of the module process.

6.1.3.2 Piping

This discipline has historically "led" the design team on stick-built projects. Again, with the shift on needs for the fabrication yard, a piping lead must also have strong leadership skills in order to be a member as they will have to help guide their discipline in "adjusting" to the slightly different requirements of a module project (and the need to support the civil/structural group).

6.1.3.3 Process

With the need to "move" equipment around to better fit within the confines of the modules, this team member needs to be open and flexible in terms of alternative arrangements for their process equipment, understanding that in some cases, minor adjustments (e.g., pump head requirements, elevation changes, etc.) may be needed to result in the best module equipment layout. They need to be able to identify such minor changes and be open to accommodating them if the result is a greatly improved module equipment and piping arrangement and the process is not compromised.

6.1.3.4 Construction

This group is always critical, as this group is the recipient of the design and fabrication efforts and, in effect, the project engineering "client." But, on the other hand, construction needs to be flexible in developing the project schedule in order to accommodate a later delivery of the modules and be proactive in communicating its needs back to the other groups.

The construction group lead needs to understand the ramifications of the module schedule and its inherent inflexibility compared to the typical and historical stick-built execution plan. As such, the construction group lead needs to follow a more collaborative approach between

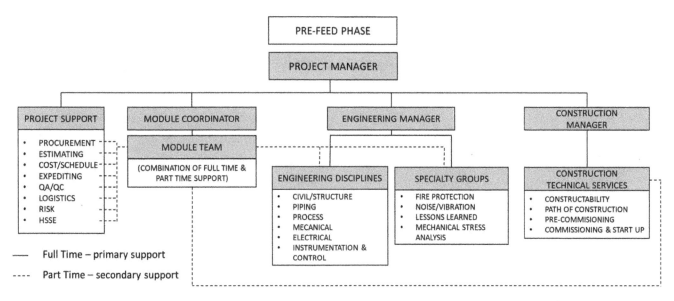

Figure 6.1 Typical pre-FEED phase module team integration into organization chart.

construction and engineering where each group must rely on the other for communication and feedback. This will ensure the move away from the past, where a somewhat more transactional approach was sometimes employed by the construction team, with the goal of just getting the materials to the site and then figuring out how best to assemble them.

6.1.3.5 Procurement

This group must adapt to changes in the procurement strategy with respect to the newer needs for "partial engineering deliverables" for early modular design.

6.1.3.6 Sub-Contracting

This group must deal with the additional segmented contracting of the fabrication yard as well as the vessels required to ship the modules.

6.1.4 Integration into Project Management

It does the project no good if the module team has a great leader and a great people-oriented module progressive team but does not properly fit the module team into the organization. The module team is a unique group that supports project management but has ties directly

to the rank and file of the various project groups. Since the team receives its input from various groups, it should report out at a management level above these groups. See Figure 6.1 for typical pre-front-end engineering design (FEED) module team integration. Figure 6.1 is the same as Figure 2.3.

Notice that the module coordinator is on the same level as the Engineering, Construction, and Project Support Services Managers, reporting directly to the project manager. This structure funnels all module-related project information directly to project management and above any other group managers who may not be as proactive in making changes required for the modular philosophy. The dotted lines are the module team representatives' secondary ties to the module coordinator and module team.

6.1.5 Module Team Growth

As the project progresses into front-end engineering design (FEED), the module team and organization expand to reflect the additional team members now working on the modules themselves, but the reporting and relationship in the project organization remain the same. See Figure 6.2 for a typical FEED and EPC module team integration. Figure 6.2 is the same as Figure 2.4.

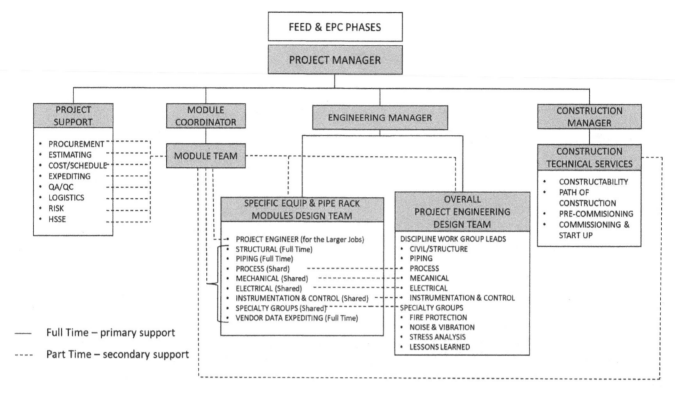

Figure 6.2 Typical FEED and EPC phases organization chart for modularization.

6.2 Execution Plan Differences (EPDs)

The second aspect of the modular philosophy we want to delve into in this chapter is how this philosophy is incorporated into the overall project planning. These are referred to as execution plan differences (EPDs). These are differences between the stick-built project execution and the modular project execution philosophies. These differences begin impacting the project as soon as the project has been identified (in opportunity framing) and continue through EPC with most of the major impacts in selection (pre-FEED) and basic design (FEED). During the research development of CII RT-283 (Construction Industry Institute, 2013), these EPDs were identified along with their approximate timing in terms of a project phase. CII RT-283 identified 107 differences covering 21 different project topics. Later, O'Connor, O'Brien, and Choi (2016) published them in an academic journal. See Table 6.1 for the 21 execution planning topics and the number of planning differences by each topic.

Rather than reiterate the information found in the references (Construction Industry Institute, 2013; O'Connor, O'Brien

and Choi, 2016) under EPDs, we will focus on some of the more prominent differences in more detail. Furthermore, we have provided our thoughts in terms of the priority and timing of each of these EPDs below. There is nothing historical or scientific in these comments, only our perception based on past modular jobs we have been associated with. Refer to (O'Connor, O'Brien, and Choi, 2016) for the full list of execution differences. The authors would like to note that many proposed EPDs below are earlier than the original EPDs identified by the Construction Industry Institute (2013) and O'Connor, O'Brien, and Choi (2016).

6.2.1 Guiding Principles

We based our "newer" ranking and earlier execution on a couple of guiding principles.

6.2.1.1 Planning Must Start Early

Even if you are not sure of the details or implementation approach, if the project can identify the need, it will go a long way in getting it addressed and solved. Most, if not

Table 6.1 The 21 execution planning topics and number of planning differences by topic.

Project phase	Topic number	Execution planning topic	Number of planning items/differences
A. Plans for selection	A-1	Project objectives	1
	A-2	Organization and staffing	4
	A-3	Health, safety, security, and environment and social impacts	3
	A-4	Craft labor relations	1
	A-5	Contract strategy	5
	A-6	Procurement strategy and owner-furnished equipment	6
	A-7	Planning and cost estimating	10
	A-8	Transport route study and planning	5
	A-9	Risk management	2
B. Plans for basic design	B-1	Stakeholder alignment and refraining	4
	B-2	Modularization scoping, layout process, and plot plan	12
	B-3	Fabricators, contractors, and subcontractors	4
	B-4	Methods, heavy lifts, and construction facilities	3
	B-5	Procurement, vendor data, and expediting	6
	B-6	Basic design standards, models, and deliverables	10
	B-7	Modularization business case validation/refinement	2
	B-8	Scope freeze and change management	5
C. Plans for engineering, procurement, and construction	C-1	Project controls and site management	7
	C-2	Quality assurance/quality control	6
	C-3	Detailed design deliverables	8
	C-4	System testing, commissioning, and start-up	3

Source: O'Connor, O'Brien and Choi (2016). Reproduced with permission of American Society of Civil Engineers (ASCE).

all, of these EPDs need to be initially addressed in earlier project phases than shown in the original Execution Plan Differences identified by the Construction Industry Institute (2013) and O'Connor, O'Brien, and Choi (2016).

6.2.1.2 Previous Module Job Experience

Because there are so many EPDs between the stick-built project and the modular project, the module project management must include the support of a module coordinator and supporting team who have previous module job experience.

6.2.1.3 Project Management Support

Again, because of all the project execution differences, the project's top management (for both the owner/client and the EPC contractor) must support the implementation of these EPDs. This is not just lip service in meetings but the actual hands-on approach to the day-to-day project activities. Project management needs to understand enough about the module execution to be able to address how deviations might impact the overall project. If they are not well versed in these differences, then project management needs to make sure that they include the module coordinator and team as required to help identify and address the impacts of deviations in the project module plan brought on by the modularization decision.

6.2.2 Topics of Execution Plan Differences

The list below provides possible EPD topics for a modular project:

• Project Objectives

• Organization & Staffing

- Modularization Business Case Validation/Refinement

- Stakeholder Alignment and Reframing

- Modularization Scoping, Layout Process, & Plot Plan

- Procurement Strategy and Owner-Furnished Equipment

- Basic Design Standards, Models, & Deliverables

- Transport Route Study & Planning

- Fabricators, Contractors, & Subcontractors

- Planning & Cost Estimating

- Craft Labor Relations

- Contract Strategy

- HSSE & Social Impacts

- Risk Management

- Methods, Heavy Lifts, & Construction Facilities

- Procurement, Vendor Data, & Expediting

- Scope Freeze & Change Management

- Project Controls & Site Management

- Quality Assurance/Quality Control

- Detailed Design Deliverables

- System Testing, Commissioning, & Start-up

6.2.3 Detailed List of Execution Plan Differences

The list below provides additional details on the EPDs for a modular project. They have been listed in order of importance (in the authors' opinion) within each project phase. They were organized in this manner to guide the project administrator or project supervisor when they are trying to determine how they should invest their available limited project planning time when hard choices on priorities will not allow every topic to be fully addressed. Also, it was assumed that, by placing the higher priority EPDs at the top, these should also receive more attention earlier in the project phases. Second, these are guidelines, not hard and fast rules. As such, there will be variations that are unique to each project. However, this should be considered a good solid starting point when determining where changes might need to occur within a project organization and its planning and when one might address them.

6.2.3.1 Opportunity Framing Phase

Project Objectives

- Objectives don't change.

- Timing will.

Organization & Staffing

- Set up a module coordinator.

- Find someone with past experience in planning module jobs.

- They will report directly to the project manager.

Modularization Scoping, Layout Process, & Plot Plan

- Identify anticipated maximum module size.

- This is critical to project success.

- Use the business case flowchart for initial solution ideas.

Transport Route Study & Planning

- Identify the route.

- Identify the maximum module shipping envelope (as accurately as you can).

Planning & Cost Estimating

- Planning: Incorporate gross benefits into the level 0 or 1 schedule.

- Cost estimating: Build project module savings into total installed cost (TIC).

Craft Labor Relations

- Identify local craft resources (or lack thereof).

Contract Strategy

- Understand the potential impact of module fabrication on contracting.

- Consider the potential for pre-FID funding options.

Risk Management

- Set up risk pros and cons for the module option.

6.2.3.2 Assessment Phase (FEL-1)

Project Objectives

• Basically complete (w/r/t module option).

Organization & Staffing

• Module team lead selected.

• Team preliminarily identified.

Modularization Business Case Validation/Refinement

• Update initial module analysis from opportunity framing business case.

• Confirm modules and cases still make sense.

Stakeholder Alignment and Reframing

• Present module business case to key stakeholders.

• Initiate initial module education of main project members.

Modularization Scoping, Layout Process, & Plot Plan

• Evaluate module options for each project option.

• Begin layout work of modules based on PFDs or existing plant information.

• Create new modules, do not "cut up and box" a similar stick-built plant.

Transport Route Study & Planning

• Commission transport feasibility study of options.

• Refine and finalize maximum transport envelope available.

Fabricators, Contractors, & Subcontractors

• Identify those that might be associated with modules.

• Identify potential fab yard candidates.

Planning & Cost Estimating

• Update planning benefits and incorporate impacts into the schedule.

• Ensure major module impacts (e.g., shipping, transport, and hook-up) are addressed.

• Update cost estimates to reflect the better definition of module scope and savings for each option.

• Determine if pre-FID funding of module design will be required.

Craft Labor Relations

• Identify local requirements.

• Investigate ability of local labor to support the project.

• Consider pros and cons of work scope move to offsite.

• Assess impacts on trade unions and local craft.

• Begin mitigation efforts on all.

Contract Strategy

• Identify contract options and impact to project module options.

• Confirm ability to include pre-FID funding options.

• Understand that modularization may shift risk balance between owner and contractor.

• Begin evaluation of early fab yard and transport contract options and strategies.

Health, Safety, Security & Environment (HSSE) & Social Impacts

• Module job provides benefits for both.

• Need to consider pros and cons of module decision and its impact on the local community.

Risk Management

• Identify benefits of better quality, craft skills, schedule surety, and lower costs.

• Compare to potential impacts due to delays from changes, transportation, late arrival, and incomplete modules.

6.2.3.3 Selection Phase (Pre-FEED or FEL-2)

Organization & Staffing

• Module team staffing increases as the project grows.

• More formal module orientations begin for project members.

Modularization Scoping, Layout Process, & Plot Plan

- Complete the modularization scoping plan.

- Plot plan, module scope, path of construction incorporated in the latest layout.

- Set module density and sizes.

- Confirm module sequencing and installation methods.

- Establish basic design philosophy for onboarding, transport, offloading, and setting.

- Ensure that construction and O&M inputs received and incorporated in design efforts,

- Installation aids considered and incorporated where appropriate.

- Module breaks incorporated into PFDs/P&IDs.

- Consider standardization where possible.

Stakeholder Alignment and Reframing

- All key stakeholders on board with module project philosophy.

- Extra efforts are required to educate new project members on module concepts as the project team grows.

- Communications critical across all interfaces.

- Team building/alignment sessions increase in size and frequency.

Fabricators, Contractors, & Subcontractors

- Screen lists of these third party companies that will support module development.

- Pre-qualify main fab yard candidates.

- Begin matching fab yard needs to current engineering design priorities and adjust.

- Determine vendor timing and match against preliminary needs from the fab yard.

- Develop timing of bulk materials to match the needs of the fab yard.

- Compare fab yard capabilities in terms of steel, piping, equipment, and bulk purchase requirements as well as the ability to perform the pre-commissioning scope.

Contract Strategy

- Complete in terms of philosophy, risk sharing, collaboration, etc.

- Maximum pre-FID funding identified for module design.

- Early fab yard and transport contract strategies are identified, and early commitments discussed.

Procurement Strategy and Owner-Furnished Equipment

- Strategy for module versus non-module vendors identified and options developed.

- Special inputs (e.g., extended warranties) worked into contracts.

- Special instructions developed for owner-furnished equipment.

- Special procedures (e.g., preventative maintenance and shipping procedures) included.

Basic Design Standards, Models, & Deliverables

- Coordination and adjustment of owner/client design standards required for a modular approach.

- O&M input solicited and included for selected project options.

- Discuss timing of detailed design development and design freeze requirements.

- 3D model developed in assessment now incorporates selected process.

- 3D model grows in detail.

- Preliminary 3D model reviews begin (in terms of gross aspects of the module design).

Transport Route Study & Planning

- Complete transport route plan.

- Route selected and maximum size identified.

- Heavy haul contractor included as part of the project team.

- Heavy lift or roll-on-roll-off vessel contracting initiated, based on module scope numbers or sizes.

Planning & Cost Estimating

- Complete planning.

- All benefits and challenges of the module option have been included in the project schedule.

- Backward pass of schedule performed to identify any problems with module deliverables.

- Ensure all module impacts (not previously identified) are built into schedule (e.g., scope carryover, logistics delays, late engineering, changes) and are addressed.

- Complete cost estimating of base case.

- Update cost estimate to reflect all benefits (e.g., softer issues difficult to assign financial benefit to) as well as challenges in terms of coordination (e.g., customs clearance, VAT) and adopt a better definition of module scope.

- Work specific needs for pre-FID funding of module design.

Craft Labor Relations

- Decisions on local labor/camps/off-site construction split completed.

- Work with local organized labor and craft resource avenues completed.

- Options on alternatives identified and vetted.

Health, Safety, Security & Environment (HSSE) & Social Impacts

- Complete HSSE impact assessment.

- Local community impacts identified and mitigated.

- Options to enhance local relations and interfacing set up for project and community.

- Communications protocol for continued two-way interfacing activated.

Risk Management

- Continue development of risk pros and cons for module option selected.

- Identify any black swan event that might apply to the project and determine if an additional examination is required from a module perspective.

- Create mitigation plans for all risks.

- Rank and follow up in next phase.

Methods, Heavy Lifts, & Construction Facilities

- Identify required heavy lifts, heavy transports within the proposed path of construction.

- Coordinate non-module equipment sets with module sequencing.

- Ensure space is available for on-site crane erection and access.

- Evaluate the need for special cranes (e.g., tower cranes) around modules.

- Ensure adequate heavy haul path through plot plan for modules.

Procurement, Vendor Data, & Expediting

- Identify purchase orders (POs) to be set up early for module needs.

- Split materials by individual module and site.

- Match delivery to module fab schedules.

- Inspections and expediting are required earlier for module scope.

- Identify instances of early purchase of engineering details for module steel design.

Scope Freeze & Change Management

- Understand the need for earlier issued for construction (IFC) documents of module design.

- Reinforce scope freeze philosophy mentioned in the assessment phase.

- "No change" mindset adopted.

- "If it works, and is safe," no change to be made.

- Strong change management program implementation required.

- Changes must pass rigorous schedule, and high project cost (ROI or NPV) hurdles.

- On-site team with authority to take and administer decisions on the spot.

Project Controls & Site Management

- Continued evolution of previous cost and schedule efforts.

- Materials and equipment tracking finalized by module and location.

- Module interfaces identified, to be confirmed in basic design.

- Confirm import and customs clearance regulations identified previously.

- Ensure construction team interfacing well with engineering and procurement.

- Develop a plan for accommodating unanticipated module scope "carryover."

Quality Assurance/Quality Control

- Complete codes and standards compliance part of fab yard qualification effort.

- Produce a plan for QA/QC team as part of the construction management team (CMT) at the fab yard.

- Develop a detailed plan for interface management and fabrication tolerance control.

- Develop vendor equipment inspection and preservation plans.

Detailed Design Deliverables

- Confirm module details, supporting fabrication, transportation, lifting, and setting at site match fab yard and site capabilities.

- Confirm design and deliverables (steel design, piping isometrics, etc.) are scheduled to complete in time to meet the fab yard start date.

System Testing, Commissioning, & Start-up

- Develop PFDs and P&IDs that reflect the path of construction as well as commissioning systems.

- Develop robust documentation procedure for sign-off of all completed pre-commissioning efforts at the fabrication yard.

- Identify and have full agreement from all parties on authority required for system sign-off.

- Ensure path of construction incorporates the most effective and efficient commissioning and start-up program possible.

6.2.3.4 Basic Design Phase (FEED or FEL-3)

Organization & Staffing

- Increase module team staffing to support the project as it grows.

- Ensure consistency of message.

Stakeholder Alignment and Reframing

- Complete stakeholder alignment and reframing.

- Project is working as a single team with respect to module decisions.

- Owner/client is on board with the execution plan.

- All stakeholders on board with module project philosophy.

- Interaction between module team and project groups is seamless.

Scope Freeze & Change Management

- Finalize scope freeze and change management.

- "No change" mindset.

- "If it works, and is safe," no change warranted.

- Strong change management program implemented.

- On-site team with authority to take and administer decisions on the spot.

Procurement Strategy and Owner-Furnished Equipment

- Complete procurement strategy.

- Module versus non-module vendors identified and contracting options firmed up.

- Special contracting language (e.g., extended warranties) incorporated.

- Owner-furnished equipment adjustments made.

- Special procedures included.

Fabricators, Contractors, & Subcontractors

- Contracting with third party companies supporting module development finalized.

- Fabrication yard selected and contracted.

- Continued fab yard and engineering coordination meetings.

- Resolve any issues with vendor equipment and materials.

- Work with fabrication yard on contingency plans for specific pieces of equipment that might be late.

- Execute design to support module fabrication.

Basic Design Standards, Models, & Deliverables

- Complete basic design standards, model, and deliverables.

- Design standards ready and in use.

- O&M input received and incorporated.

- 3D model under full development.

- Model reviews are scheduled on a regular basis.

Methods, Heavy Lifts, & Construction Facilities

- Complete review of methods, heavy lifts, and construction facilities.

- Contract of heavy lift with heavy transport companies in place.

- Path of construction confirmed; all equipment and modules to be set as planned.

- Space available to assemble and set everything is confirmed.

- Special cranes (e.g., tower cranes) to support final piping placement around modules after setting contracted.

- Heavy haul path details through plant completed.

Procurement, Vendor Data, & Expediting

- Complete review of procurement, vendor data, and expediting.

- Module purchase orders (POs) placed and prioritized.

- All materials and equipment identified by module and destination.

- Equipment delivery accommodates module fab schedules.

- Early module inspections and expediting identified and set up.

- Early vendor data requirements for module design incorporated into contracts.

Project Controls & Site Management

- Complete project controls and site management.

- Cost and schedule efforts finalized.

- Actively monitor materials and equipment tracking.

- Module interfaces confirmed.

- Import and customs clearance working smoothly.

- Construction team interfacing well with fabrication yard.

- Everyone monitoring progress to eliminate module scope "carryover" to site.

Quality Assurance/Quality Control

- Implementing inspections at equipment vendors.

- Working with fab yard to coordinate inspections when they begin fabrication.

- Working with all parties to complete sign-off procedures for pre-commissioning efforts at fab yard.

- Detailed plan for interface management and fabrication tolerance control in place.

- Finalized single weld hook-up (SWHU) procedures for piping between modules.

Detailed Design Deliverables

- Completed design of module details, supporting fabrication, transportation, lifting, and setting criteria at site match requirements of fab yard and site capabilities.

- Completed in time to support fab yard start date.

System Testing, Commissioning, & Start-up

- Execute design based on systems identified on the PFDs and P&IDs that reflect path of construction and commissioning system requirements.

- Develop test packs for all pre-commissioning efforts to be completed at the fabrication yard and signed off.

Table 6.2 Suggested implementing timing sequence of the execution plan topics.

Modular execution plan topics	Suggested optimal timing and sequencing				
	Opportunity framing	Assessment	Selection	Basic design	EPC
Project Objectives	1	1-complete			
Organization & Staffing	2	2-complete	increase	increase	increase
Modularization Business Case Validation/Refinement		3	1	1-complete	
Stakeholder Alignment and Reframing		4	3	2-complete	
Modularization Scoping, Layout Process, & Plot Plan	3	5	2-complete		
Procurement Strategy and Owner-Furnished Equipment			6	4-complete	
Basic Design Standards, Models, & Deliverables			7	6-complete	
Transport Route Study & Planning	4	6	8-complete	confirm	
Fabricators, Contractors, & Subcontractors		7	4	5-complete	
Planning & Cost Estimating	5	8	9-complete	update	confirm
Craft Labor Relations	6	9	10-complete		
Contract Strategy	7	10	5-complete		
HSSE & Social Impacts		11	11-complete		
Risk Management	8	12	12	13-complete	
Methods, Heavy Lifts, & Construction Facilities			13	7-complete	
Procurement, Vendor Data, & Expediting			14	8-complete	
Scope Freeze & Change Management			15	3-complete	
Project Controls & Site Management			16	9-complete	
Quality Assurance/Quality Control			17	10-complete	
Detailed Design Deliverables			18	11-complete	
System Testing, Commissioning, & Start-up			19	12-complete	

Notes: 1, 2, 3, etc. = priority within project phase; 1-complete = priority within project phase and activity completed.

6.2.3.5 The EPC Phase

• Execute basic design package.

• Avoid changes.

• Monitor progress.

• Adjust as required.

6.2.4 Timing of Execution Plan Differences

The above-detailed listing of the EPDs is summarized in Table 6.2. As noted earlier, the key differences between the original EPDs identified by the Construction Industry Institute (2013) and O'Connor, O'Brien, and Choi (2016) and the currently proposed EPDs shown in Table 6.2 are ones of timing, with most shifting to the left (to an earlier phase for both initiation as well as completion).

In considering the approach to these for a given project, examine them in terms of importance as well as timing. For each phase of the project, it is best to evaluate these EPDs in an orderly and logical sequence.

Not every EPD can be initiated in the earliest phases of a project, but every EPD needs to be considered as early as possible and especially when supporting information has become available. Just as the business case flowchart is an iterative process and the results from one iteration become the inputs to the subsequent iteration, a similar evolution occurs with the development of the EPDs. At the start of working on an activity in the early project phases, you probably will not have either all the information or knowledge needed to complete it, so the activity must mature over two or more project phases. As you can see in Table 6.2, initially, the lower priority items in one project phase become a higher priority in the next project phase, and as they are finally completed, they fall off the chart.

6.3 Summary

It is hoped that this chapter on the modular team and explanation of the EPDs will, at a minimum, provide a springboard for organizing your project if you are unfamiliar with module coordination and module execution differences, both obvious and subtle.

In terms of the modular team, it is critical that both the owner/client as well as the pre-FEED, FEED, and EPC contractors have module experience, that module teams have been identified, and they begin talking to each other at the earliest possible opportunity. One driving requirement for the success of a modular project execution is that the entire team moves along the same path to the same goals. This is best orchestrated as a defined team with that as its only goal.

For those readers with some previous module planning experience, you may disagree with some of the priorities as well as the timing. As mentioned, this listing will vary from one project to the next, and priorities will need to be adjusted to fit what is considered critical for that project. But even if you don't agree with everything as listed in Table 6.2 and the list of EPDs, it should provide a goal for implementation as well as a tickler file of potential activities to address. Use both to help flesh out details of your project when planning it. You can't go wrong with too much planning.

References

Construction Industry Institute (2013) *Industrial Modularization: Five Solution Elements*. Austin, TX: The University of Texas at Austin: Construction Industry Institute.

O'Connor, J.T., O'Brien, W.J., and Choi, J.O. (2016) Industrial Project Execution Planning: Modularization versus Stick-Built. *Practice Periodical on Structural Design and Construction*, 21(1), 04015014. doi: 10.1061/(ASCE) SC.1943-5576.0000270.

Key Critical Success Factors for Modular Project Success

This chapter introduces the critical success factors (CSFs) that contribute the most to the modular project performances. CSFs are common key elements of successful modular projects. Modular project job failures are directly linked to the failure to follow one or more of these CSFs. Conversely, modular project job successes address most, if not all, of these CSFs.

The authors, with multiple colleagues, have been investigating this topic for a long time. First, Construction Industry Institute (CII) RT-283 (Construction Industry Institute, 2013), where the authors participated as research members, identified 21 key CSFs and enablers for modular project success. CII (2013) provided the list of CSFs and the responsible party and the recommended timing of CSFs. Later, as these CSFs were mainly found based on the subject matter experts' opinion, there was a need to further investigate these CSFs with actual modular projects. Thus, Choi (2014) identified links between these CSFs and modular project performances and investigated the CSFs' accomplishment status. Later, Choi, O'Connor, and Kim (2016) identified cost and schedule success recipes for modular plants.

This chapter would be lengthy if the authors repeated or included all the findings on modularization CSFs. Instead, the authors present only the essence of knowledge supported by multiple research studies (Construction Industry Institute, 2013; Choi, 2014; Choi and O'Connor, 2014; O'Connor, O'Brien, and Choi, 2014; Choi, O'Connor, and Kim, 2016; Choi *et al.*, 2019).

First, this chapter introduces a complete list of 21 CSFs with their labels and description. Second, the authors provide the frequent module job mistakes and the least

accomplished CSFs the industry needs to pay attention to. Third, responsible parties and recommended timings for each CSF are provided. The authors also highlight the most delayed CSFs in terms of accomplishment timing, which the industry needs to accomplish early. In addition, the authors explain the importance of implementing CSFs inclusively by showing associations between modularization CSFs and project performance and the recipes for cost and schedule success for modular industrial projects. Finally, the chapter concludes with a training exercise that challenges the reader through a plausible story, a module we like to call "the Perfect Storm."

7.1 Modularization Critical Success Factors (CSFs)

To identify the key modularization CSFs, CII RT-283 (2013) identified 72 potential success factors and ranked them in terms of significance by using a survey of modularization subject matter experts. Consequently, 21 key modularization CSFs were selected, which were of higher significance. The complete list of 72 potential success factors can be found in (O'Connor, O'Brien, and Choi, 2014).

7.1.1 CSF Labels and Descriptions

A complete list of the most significant 21 CSFs with their labels and descriptions are as follows (O'Connor, O'Brien, and Choi, 2014), reproduced with permission of American Society of Civil Engineers (ASCE):

- CSF1 MODULE ENVELOPE LIMITATIONS: Preliminary transportation evaluation should result in understanding module envelope limitations.

- CSF2 ALIGNMENT ON DRIVERS: Owner, consultants, and critical stakeholders should be aligned on important project drivers as early as possible in order to establish the foundation for a modular approach.

- CSF3 OWNER'S PLANNING RESOURCES & PROCESSES: As a potentially viable option to conventional stick-building, early modular feasibility analysis is supported by owner's front-end planning and decision support systems, work processes, and team resources support. Owner's 'comfort zones' are not limited to the stick-built approach.

- CSF4 TIMELY DESIGN FREEZE: Owner and contractor are disciplined enough to effectively implement timely staged design freezes so that modularization can proceed as planned.

- CSF5 EARLY COMPLETION RECOGNITION: Modularization business cases should recognize and incorporate the economic benefits from early project completion that result from modularization and those resulting from minimal site presence and reduction of risk of schedule overrun.

- CSF6 PRELIMINARY MODULE DEFINITION: Front-end planners and designers need to know how to effectively define the scope of modules in a timely fashion.

- CSF7 OWNER-FURNISHED/LONG-LEAD EQUIPMENT SPECIFICATION: Owner-furnished and long-lead equipment (OFE) specification and delivery lead-time should support a modular approach.

- CSF8 COST SAVINGS RECOGNITION: The modularization business case should incorporate all cost savings that can accrue from the modular approach. Project teams should avoid the knee-jerk misperception that modularization always has a net cost increase.

- CSF9 CONTRACTOR LEADERSHIP: Front-end contractor(s) should be proactive—supporting the modular approach on a timely basis and prompting owner support, when owner has yet to initiate.

- CSF10 CONTRACTOR EXPERIENCE: Contractors (supporting all phases) have sufficient previous project experience with the modular approach.

- CSF11 MODULE FABRICATOR CAPABILITY: Available, well-equipped module-fabricators have adequate craft, skilled in high-quality/tight-tolerance modular fabrication.

- CSF12 INVESTMENT IN STUDIES: In order to capture the full benefit, owner should be willing to invest in early studies into modularization opportunities.

- CSF13 HEAVY LIFT/SITE TRANSPORT CAPABILITIES: Necessary heavy lift/site transport equipment and associated planning/execution skills are available and cost-competitive.

- CSF14 VENDOR INVOLVEMENT: OEMs [original equipment manufacturers] and technology partners need to be integrated into the modularized solution process in order to maximize related beneficial opportunities.

- CSF15 OPERATIONS & MAINTENANCE (O&M) PROVISIONS: Module detailed designs should incorporate and maintain established O&M space/access needs.

- CSF16 TRANSPORT INFRASTRUCTURE: Needed local transport infrastructure is available or can be upgraded/modified in a timely fashion while remaining cost-competitive.

- CSF17 OWNER DELAY AVOIDANCE: Owner has sufficient resources and discipline to be able to avoid delays in commitments on commercial contracts, technical scope, and finance matters.

- CSF18 DATA FOR OPTIMIZATION: Owner and pre-FEED/FEED contractor(s) need to have management tools/data to determine the optimal extent of modularization, i.e., maximum net present value (NPV; that considers early revenue streams) vs. percentage modularization.

- CSF19 CONTINUITY THROUGH PROJECT PHASES: Disconnects should be avoided in any contractual transition between assessment, selection, basic design, or detailed design phases; their impacts can be amplified with modularization.

- CSF20 MANAGEMENT OF EXECUTION RISKS: Project risk managers need to be prepared to deal with any risks shifted from the field to engineering/procurement functions.

- CSF21 TRANSPORT DELAY AVOIDANCE: Environmental factors such as hurricanes, frozen seas, or lack of permafrost, in conjunction with fabrication shop schedules, do not result in any significant project delay.

A failure to follow these CSFs leads to a disastrous result and erroneous conclusion about modularization. The CII (2013) further provided CSF enablers that can facilitate the implementation of each CSF. The authors recommend the reader to check them out from (Construction Industry Institute, 2013) or (O'Connor, O'Brien, and Choi, 2014).

7.1.2 Frequent Module Job Mistakes

The authors have witnessed many mistakes (in other words, failure to implement CSFs) made in the actual modular projects. Here are some of the potential module job mistakes made. The number in brackets refers to the CSF list above.

- Incomplete Module Team (15)
  - Construction—primary customer—not included
  - O&M—ultimate customer—input too little or too late
- Poor planning (2, 3, 6)
  - Module analysis too late
  - Inadequate time to develop process and incorporate
  - Failure to "do it differently"
- Failure to commit necessary (pre-financial investment decision, FID) funding (12, 17)
  - Late engineering deliverables
  - Missing critical equipment design data
  - Poor module fabrication strategy
- Poor engineering execution (7, 14, 18)
  - Engineering not started early enough—impacts deliverables—issued for construction (IFC) drawings
  - Incomplete design provided to fab yard
  - Assuming change flexibility "at the fab yard" is acceptable
  - Failure to deliver bulks and equipment on time
- Starting fabrication too early (19, 20)
  - Fab yard runs out of work

- Options: Expensive expediting or expensive fab yard delays
- Late engineering changes (4, 6)
  - Disrupts fab yard sequence
  - Requires extensive changes to completed structure
- Shipping module incomplete (21)
  - Defeats decision to use fab yard
  - Site not equipped to pick up extra scope
- Poor coordination between fab yard and site (9, 10, 11)
  - Incomplete pre-commissioning & sign off
  - Incomplete carry over documentation
  - Incomplete. . . . . ,period

In addition, according to (Choi, 2014), the actual modular projects had difficulty in terms of accomplishing the following CSF (Figure 7.1):

- CSF7. Owner-furnished/Long Lead Equipment
- CSF8. Cost Saving Recognition
- CSF12. Investment in Studies
- CSF15. O&M Provisions
- CSF18. Data for Optimization

The industry needs to be careful not to make mistakes and implement the CSFs above more. Check (Choi, 2014) for the details.

7.1.3 Responsibility and Timing of CSFs

Acknowledging and implementing the 21 key CSFs by themselves do not guarantee a project's success. There must be a correct timing of implementation by the acknowledged lead party responsible for ensuring success. Table 7.1 lists the optimal implementation timing and responsible party for each CSF.

What is interesting about the 21 CSFs when they are analyzed in terms of who is responsible as well as when the CSF should be implemented are the following:

- More than half (57%) are the direct responsibility of the client or owner, not the EPC contractor.

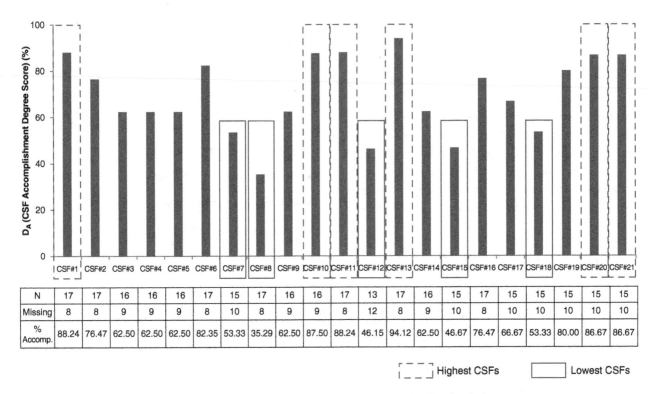

	CSF#1	CSF#2	CSF#3	CSF#4	CSF#5	CSF#6	CSF#7	CSF#8	CSF#9	CSF#10	CSF#11	CSF#12	CSF#13	CSF#14	CSF#15	CSF#16	CSF#17	CSF#18	CSF#19	CSF#20	CSF#21
N	17	17	16	16	16	17	15	17	16	16	17	13	17	16	15	17	15	15	15	15	15
Missing	8	8	9	9	9	8	10	8	9	9	8	12	8	9	10	8	10	10	10	10	10
% Accomp.	88.24	76.47	62.50	62.50	62.50	82.35	53.33	35.29	62.50	87.50	88.24	46.15	94.12	62.50	46.67	76.47	66.67	53.33	80.00	86.67	86.67

┌ ─ ─ ┐ Highest CSFs   ☐ Lowest CSFs

Figure 7.1   Degree of CSFs accomplishment by each CSF. Source: Choi (2014). Reproduced with permission of Jin Ouk Choi.

Table 7.1   CSFs' optimal timing and responsible party.

	Optimal timing	Responsibility
CSF#1	Starting in Assessment	Owner
CSF#2	Opportunity Framing and each subsequent phase	Owner
CSF#3	Opportunity Framing through Selection	Owner
CSF#4	Mid-Basic Design through Early EPC	Owner and Contractor
CSF#5	Assessment through Selection	Owner
CSF#6	Starting in Assessment	Owner or Contractor(s)
CSF#7	Starting in Assessment	Owner
CSF#8	Starting in Assessment	Owner
CSF#9	Starting in Assessment	Contractor(s)
CSF#10	Assessment thru Detailed Design	Contractors
CSF#11	Starting in Basic Design	Owner and Contractor(s)
CSF#12	Assessment and Selection	Owner
CSF#13	Basic Design	EPC Contractor
CSF#14	Starting in Assessment	Owner, Major Vendors, and Technology Licensor
CSF#15	Starting in Basic Design	Owner Operations and Maintenance and Contractor(s)
CSF#16	Prior to Initiation through Assessment	Owner
CSF#17	No later than Early Selection	Owner
CSF#18	Early in Selection	Owner and Contractor(s)
CSF#19	All Phases	Owner
CSF#20	Starting in Basic Design	All Parties
CSF#21	Starting in Basic Design	Contractor(s)

Source: Adopted from O'Connor, O'Brien, and Choi (2014). Reproduced with permission of American Society of Civil Engineers (ASCE).

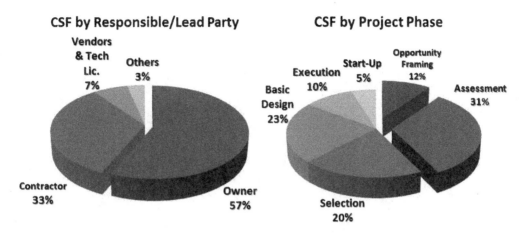

Figure 7.2 CSFs distribution pie charts by responsible/lead party and implementation phase. Source: O'Connor, O'Brien, and Choi (2014). Reproduced with permission of American Society of Civil Engineers (ASCE).

- More than 1/3 (42%) require implementation by the Assessment Phase (FEL 1 or pre-pre-FEED).

- Over half (62%) must be implemented before Basic Design (FEL 3 or FEED).

Graphically, this information is shown in Figure 7.2.

7.1.4 Most Delayed CSFs in Terms of Accomplishment Timing

Choi (2014) also investigated the status of modularization CSFs' implementation in terms of timing compared to the Construction Industry Institute's (2013) recommended timing. The most delayed CSFs in terms of accomplishment timing include the following (see Figure 7.3):

- CSF14. Vendor Involvement

- CSF16. Transport Infrastructure

- CSF1. Module Envelope Limitations

- CSF5. Early Completion Recognition

- CSF6. Preliminary Module Definition

- CSF8. Cost Saving Recognition

The industry needs to pay attention to these six most delayed CSFs. These CSFs should be implemented earlier or on correct/recommended timing to ensure module project success.

7.2 Association between Modularization CSF and Project Performance

Choi (2014) identified correlations between modularization CSFs and project performance from actual industrial modular projects to help the industry accomplish better modular project performance. This study confirmed that modular project performance is associated with modularization CSF accomplishment. In other words, the actual modular projects with a higher degree of CSF accomplishment tend to have better cost, schedule, construction, and start-up performances. The relationship between the degree of modularization CSF accomplishment and cost performance is shown in Figure 7.4, the relationship between the degree of modularization CSF accomplishment and schedule performance is shown in Figure 7.5, the relationship between the degree of modularization CSF accomplishment and construction performance is shown in Figure 7.6, and the relationship between the degree of modularization CSF accomplishment and start-up performance is shown in Figure 7.7.

These research findings highlight the importance of implementing all the CSFs, as these CSFs can impact overall project performance, including cost, schedule, construction, and start-up. Another study (Choi, O'Connor, and Kim, 2016) also found that modularization CSFs collectively affect modular project performance. Thus, the authors recommend the reader/industry should implement all the key CSFs for their modular project. If the reader feels that 21 CSFs are too many, the authors recommend paying attention to timely design freeze, owner furnished/long-lead equipment specification, vendor involvement, and management of execution risks, as these CSFs are critical for cost and schedule successes in industrial modular projects (Choi, O'Connor, and Kim, 2016).

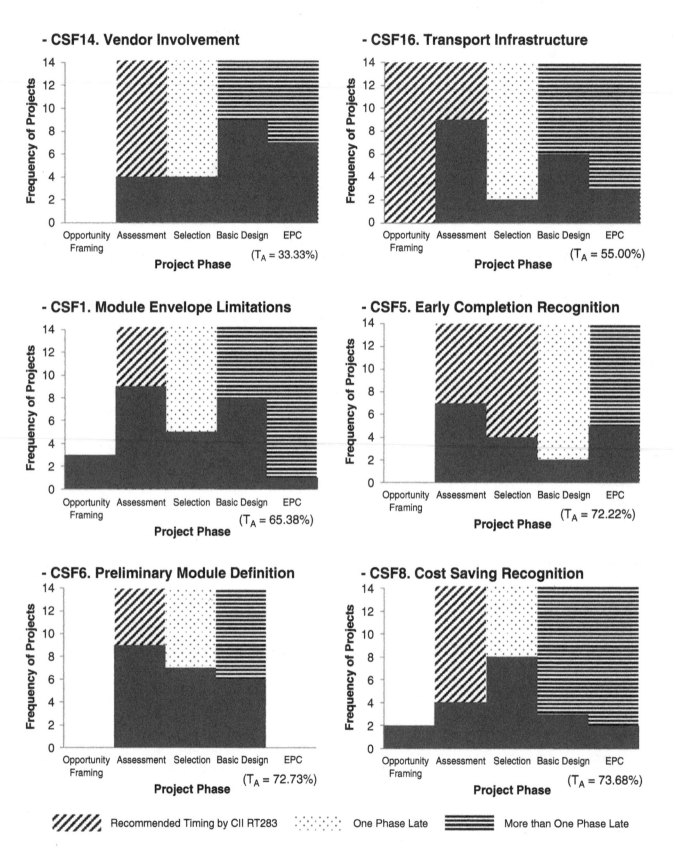

Figure 7.3 Most delayed CSFs in terms of accomplishment timing. Source: Choi (2014). Reproduced with permission of Jin Ouk Choi.

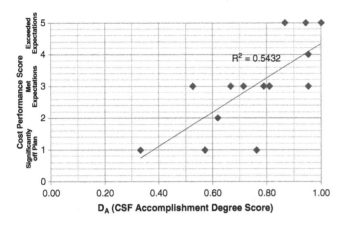

Figure 7.4 Association between modularization CSFs and cost performance. Source: Choi (2014). Reproduced with permission of Jin Ouk Choi.

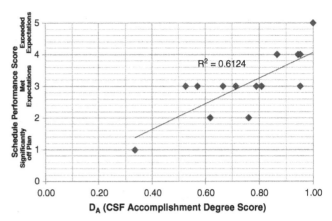

Figure 7.5 Association between modularization CSFs and schedule performance. Source: Choi (2014). Reproduced with permission of Jin Ouk Choi.

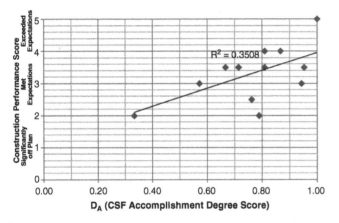

Figure 7.6 Association between modularization CSFs and construction performance. Source: Choi (2014). Reproduced with permission of Jin Ouk Choi.

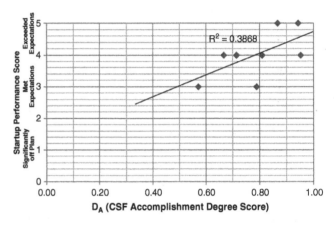

Figure 7.7 Association between modularization CSFs and start-up performance. Source: Choi (2014). Reproduced with permission of Jin Ouk Choi.

7.3 Training Exercise

We think it would be a much better explanation of what a CSF is if we could put it into the context of a plausible story, the module we like to call "the Perfect Storm." While this short story is not taken from one particular project, it is similar to many modular jobs that, for one reason or another, did not end well. However, in this case, like the movie of 2000 with Mark Wahlberg and George Clooney, where three weather fronts converge to create the Perfect Storm, our example is where the failure to implement many of the critical success factors resulted in a modular project disaster of epic proportions.

7.3.1 Instructions for the Training Exercise

While reading the module "the Perfect Storm" story, the readers/students are asked: (1) to identify the failures to implement the critical success factors, and (2) to name

the CSF number or label that should have been implemented. Tip: Not all the paragraphs have failures to implement the critical success factors, but many of them do include the failures. Some of the failures/stories may require multiple CSFs to be implemented.

7.3.1.1 The Module "the Perfect Storm"

1. The job is awarded. As always, when one wins a lump sum job, there is the nagging fear that you left something major out of your bid. Because times are tight and jobs are hard to get, the EPC company accepted the job knowing that the schedule (which is critical and tied to liquidated damages if missed) was too short.

2. The construction team immediately balked at accelerating the schedule, providing all kinds of reasons for not being able to compress their durations.

So, engineering was pushed to reduce BOTH their schedule and the work-hours needed to get the early engineering off to a good start.

3. With a less than stellar engineering effort, the equipment vendors were pushed by engineering to provide details for the equipment designs without some of the process details finalized. As a result, the data provided by the equipment vendors was either incomplete or late (or both).

4. With details necessary for a good module fabrication bid missing, the scope of work (SOW) provided to the fabrication yard was incomplete with critical equipment due dates identified by the vendors based on preliminary information and timing. The module fabrication yard bases its pricing and schedule on this optimistic delivery timing. As a result, the fabrication yard begins the module fabrication effort too early, relying on incomplete design and layouts.

5. These are quickly followed by changes to the initial solicitation from more recent equipment vendor data. More changes occur as they are incorporated into the issued for construction (IFC) drawings. This is further compounded by the late delivery of the equipment and the wrong mix of early piping bulks, resulting in inefficiencies in the assembly and spooling efforts.

6. Since the module fabrication yard contract is defined by early (and inaccurate) equipment and piping deliveries, the module fab yard begins issuing change orders for the new work/re-work/demolition work required to correct the early efforts to match the new IFC drawings. This all creates more work for the fab yard, which is more costly, but, more importantly, extends the duration of the module completion effort by the fab yard.

7. However, because of the intense push to meet the schedule, the large ocean-going roll-on-roll-off (RO-RO) vessels were contracted to arrive at the initially planned fab yard completion date. So, these large, expensive vessels arrive on time. Because these large vessels are contracted years in advance, they are dedicated to carrying specific modules. As a result, there are only two options—either (1) accept some charges for vessel contract cancelation and try to contract a second vessel that will not be available until many months or even years later, or (2) ship the incomplete module to the job site. Since the delay of months waiting for a new ocean-going vessel is not

an acceptable option in terms of the liquidated damage potential from the project contract, the decision is made to ship the incomplete module directly to the job site where it will be completed.

8. Unfortunately, the on-site workforce is not equipped with either the correct number or type of skilled workers to take on this additional work. In addition, the actual amount of carryover work being shipped to the site is not accurately identified by the fab yard as this was still a work in progress.

9. In addition, the missing equipment, as well as piping, instrumentation, and electrical bulks currently at the fab yard are now required at the site, creating a logistics challenge. The actual carryover work details have not been completed, so the actual carryover work scope is a bit of a mystery to the project site team. Also, there is a reluctance to ship anything to the project site from the fab yard, as the fab yard is still in the midst of working the rest of the module pipe spools and EI&C. The result is that the project site crew is saddled with an increase in work scope, but without all of the materials to complete the carryover work.

10. When the equipment and bulks are finally shipped from the fab yard, they are installed late as out of sequence work: a very costly and prohibitive effort. Later, it was discovered that module detailed designs did not incorporate all the operations and maintenance (O&M) space and access needs, which led to re-work at the project site.

11. Also, the large amount of pre-commissioning that was to have occurred on these incomplete modules was not completed at the fab yard. This extra pre-commissioning work must be added to the on-site construction team scope. AND, because the little bit of pre-commissioning that was completed at the fab yard was not properly documented and signed off, ALL of the pre-commissioning work completed at the fab yard is now re-worked at the site.

All during this, the schedule continues to slip to the right, and the cost overruns continue to pour in. The project site crew, which is much more expensive, must continue to increase to try to mitigate the schedule delays. The project management focuses on the module fab yard as the reason for the cost increases and delays since it is this fab yard that has been developing all the change orders and cost deviation. The fab yard is also seen as the prime reason that the modules were late and incomplete. The second-guessing begins on what went wrong and what could have been done better, but with all the focus on the module fab yard, the inevitable conclusion is that: *Modularization was a bad decision*. This conclusion is carried on to the next several subsequent projects, and the benefits of modularization are lost on these also.

7.3.1.2 What They Got Right

Despite all the bad decisions and errors in judgment, there were a few critical success factors that "the Perfect Storm" project followed. Why is it important to separately note and identify the CSFs that were successfully incorporated in "the Perfect Storm" project? Because as inferred in the more in-depth analysis of the CSFs above, at some point in the project, all the CSFs can be considered essential, so the goal is to make sure that the project understands, considers, and incorporates them all. Incorporating most of the CSFs may not be enough.

As shown by "the Perfect Storm" project, just getting a few incorporated into the project planning (even if separately, each one was considered important) was not nearly enough to make the project successful. By the same token, getting all but one of the CSFs incorporated may also still be a disaster. Does this sound a bit far-fetched? Do not think that such a scenario might never happen. Unfortunately, both authors are familiar with practical examples of where these bad scenarios have actually occurred. So, hopefully, we have made the point with our simple project example and follow-up comments that the best approach to module project success is to incorporate ALL the CSFs. Traveling down the project planning paths that either incorporate most, but not all, or incorporate only a few of the seemingly important ones will most likely end up in a module project disaster—maybe not of this epic proportion, but still one that will most likely be very unacceptable to all involved.

References

Choi, J.O. (2014) Links between Modularization Critical Success Factors and Project Performance. Ph.D. dissertation. The University of Texas at Austin. Available at: https://repositories.lib.utexas.edu/handle/2152/25030.

Choi, J.O. and O'Connor, J.T. (2014) Modularization Critical Success Factors Accomplishment: Learning from Case Studies. In *Construction Research Congress 2014*. Atlanta, GA: American Society of Civil Engineers, pp. 1636–1645. doi: 10.1061/9780784413517.167.

Choi, J.O., O'Connor, J.T., and Kim, T.W. (2016) Recipes for Cost and Schedule Successes in Industrial Modular Projects: Qualitative Comparative Analysis. *Journal of Construction Engineering and Management*, 142(10), 04016055. doi: 10.1061/(ASCE)CO.1943-7862.0001171.

Choi, J.O. *et al.* (2019) Calibrating CII RT283's Modularization Critical Success Factor Accomplishments. In Al-Hussein, M. (Ed.) *2019 Modular and Offsite Construction (MOC) Summit*. Banff, Alberta, Canada: University of Alberta, pp. 235–242. doi: 10.29173/MOCS99.

Construction Industry Institute (2013) *Industrial Modularization: Five Solution Elements*. Austin, TX: The University of Texas at Austin: Construction Industry Institute.

O'Connor, J.T., O'Brien, W.J., and Choi, J.O. (2014) Critical Success Factors and Enablers for Optimum and Maximum Industrial Modularization. *Journal of Construction Engineering and Management*, 140(6), 04014012. doi: 10.1061/(ASCE)CO.1943-7862.0000842.

chapter **8** The Fabrication Yard

The fabrication yard is the central player in the modular project, and a lot of activity revolves around it. As this player is one of the key topics to understand modularization, we have decided to cover this in a separate chapter. So, let's go over some basics about the "generic" fabrication yard, including the benefits and characteristics, selecting the right fabrication yard, the contracting strategy, and the division of responsibility. Of course, we have already addressed many of the basics in the previous chapters, so we will initially fill in the gaps in this chapter.

8.1 Basic Benefits of the Fab Yard

In Chapter 1, we have already highlighted the reasons for moving work to a fabrication yard—a dedicated facility with ample crane capacity, located in a temperate climate, and populated with a motivated and highly skilled labor workforce that works efficiently together building these modules at grade for eventual stacking. But why is the typical fabrication yard so "good?" If you take a look at the fabrication yard business model, you will see a lot of similarities to a manufacturing plant. In fact, we like to compare the module fab yard to an automobile manufacturing facility, like Ford. (only because we are partial to cars and there is so much history in the development of the assembly line, including Deming philosophy and just-in-time delivery concepts).

The typical fabrication yard has been designed to produce quickly and efficiently (and in volume). Whether it is ships, modules, or offshore platforms, it has been laid out with that end in mind. The fabrication yard develops its assembly process based on its availability of lifting equipment.

In many cases, the larger fabrication yards (in South Korea and China) have built their own specialized lifting equipment (gantry cranes) to specifically support this assembly.

To support the delivery of the completed module, there is a purpose-built wharf and quayside. To connect the assembly area(s) to the quayside is a reinforced wharf, sometimes further outfitted with a set of reinforced skid beams or skidways providing preset paths that withstand very high bearing pressures created by the movement of the extremely large structural offshore jackets. Most of the modules are designed to be supported by enough axle lines of rolling self-propelled modular transporters (SPMTs) that the typical wharf deck can handle the bearing pressures.

With this assembly line concept for the modules, two major remaining activities that need to be coordinated are the structural steel fabrication and the piping spool production. This is where the fabrication yard philosophy may deviate between the large fabrication yard and the smaller "assembly" yard. The large fabrication yard will set up "manufacturing" lines for the structural steel as well as the process piping. These areas are set up with an assembly line process in mind, with raw materials (plate steel, structural shapes or pipe and fittings) coming in at one end and stations set up for performing the cutting, rolling, tacking, bending, fit-up and welding, with the finished product coming out at the other end, to be moved to the blasting and protective coating or galvanizing areas. These areas are further divided to concentrate on small or large bore as well as carbon, stainless steel, or alloy piping. These can be as simple as an in-line process or as complicated as a multi-path assembly scheme, as shown in Figure 8.1.

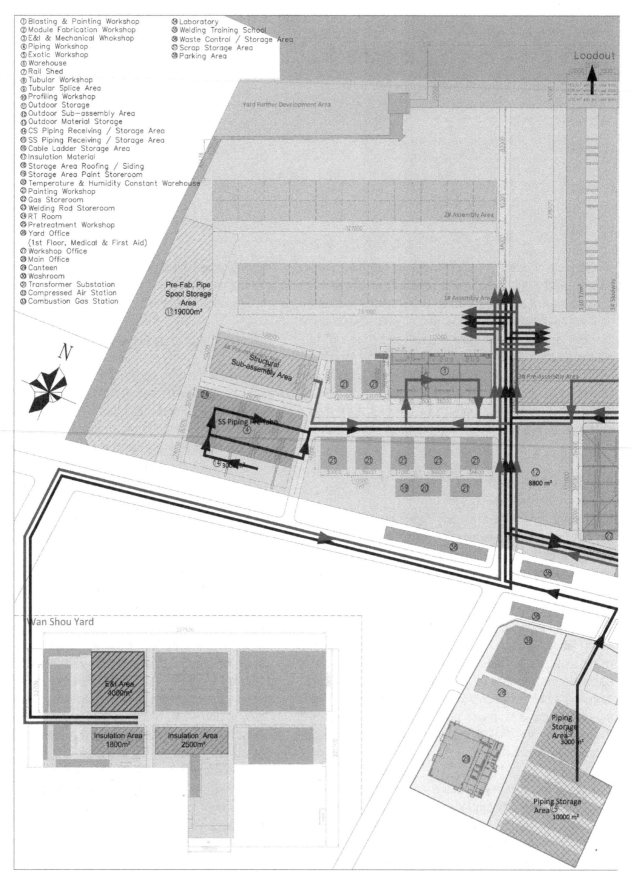

① Blasting & Painting Workshop
② Module Fabrication Workshop
③ E&I & Mechanical Whokshop
④ Piping Workshop
⑤ Exotic Workshop
⑥ Warehouse
⑦ Rail Shed
⑧ Tubular Workshop
⑨ Tubular Splice Area
⑩ Profiling Workshop
⑪ Outdoor Storage
⑫ Outdoor Sub-assembly Area
⑬ Outdoor Material Storage
⑭ CS Piping Receiving / Storage Area
⑮ SS Piping Receiving / Storage Area
⑯ Cable Ladder Storage Area
⑰ Insulation Material
⑱ Storage Area Roofing / Siding
⑲ Storage Area Paint Storeroom
⑳ Temperature & Humidity Constant Warehouse
㉑ Painting Workshop
㉒ Gas Storeroom
㉓ Welding Rod Storeroom
㉔ RT Room
㉕ Pretreatment Workshop
㉖ Yard Office
    (1st Floor, Medical & First Aid)
㉗ Workshop Office
㉘ Main Office
㉙ Canteen
㉚ Washroom
㉛ Transformer Substation
㉜ Compressed Air Station
㉝ Combustion Gas Station

㉞ Laboratory
㉟ Welding Training School
㊱ Waste Control / Storage Area
㊲ Scrap Storage Area
㊳ Parking Area

Figure 8.1 Module fabrication yard showing paths of construction. Source: Courtesy of PJOE fabrication yard.

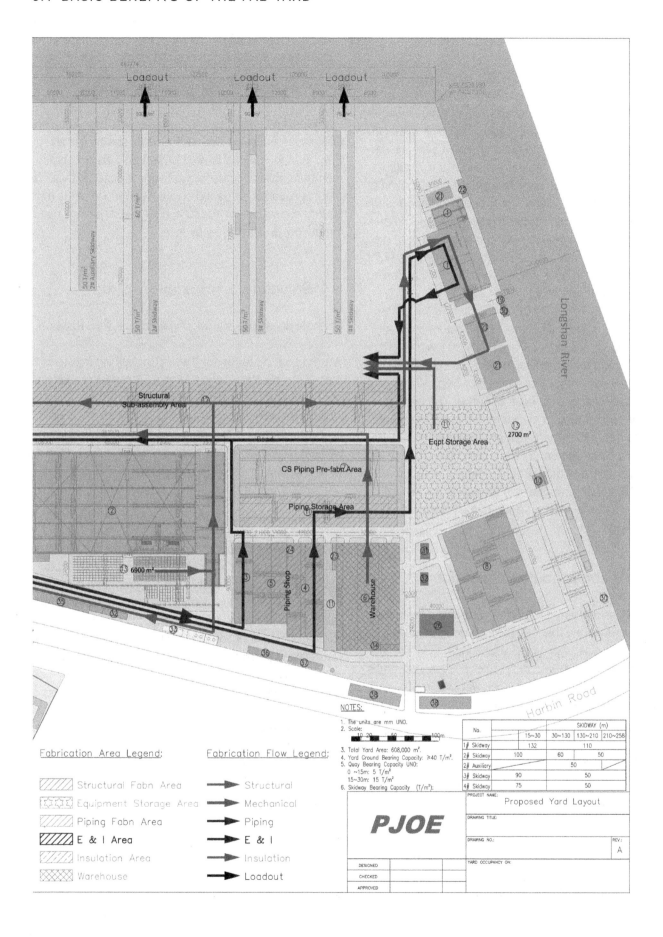

8.2 Manufacturing and Construction Mixture

The typical fabrication yard is a combination of a manufacturing plant (structural steel and pipe spools) as well as a project site (where the steel, piping, equipment, electrical and instrumentation and controls (E&I&C) are assembled) similar to a small project job site. As a result, the fabrication yard can end up with a unique combination of planning features developed from both advanced work packaging (AWP) as well as Lean processes. This creates a problem for the owner/client and the engineering, procurement, and construction (EPC) companies who want to be prescriptive in how progress and planning are being developed and monitored.

The fabrication yard (like the Ford Motor company's assembly line example) has a system that works. They are not interested in modifying it just to match some progress measurement or planning tools (e.g., AWP) that one of these clients may have adopted. The best solution where these tools may not match is for the owner/client and the EPC to discuss these issues with the fabrication yard management to understand how they work. Then, with the fabrication yard, identify outputs from their manufacturing progression tools that can be downloaded for external input into the owner/client and EPC progress tools.

8.3 AWP and Module Fabrication

That being said, many fabrication yards have embraced the AWP philosophy that has recently become so popular. And with some slight nomenclature changes, the fabrication yards have developed project planning schemes that look very similar to and work well within the AWP context, as the examples shown below:

• The construction work area (CWA) is identical to a module.

  • CWA = Module (or group of modules).

• The construction work package (CWP) is identical to the module fabrication work package (FWP).

  • CWP = FWP.

• The site installation work package (IWP) must be split into: (1) a module installation work package (for prefab); and (2) an installation work packages (for assembly).

This is because an efficient prefabrication priority is not identical to an efficient module assembly priority.

  • IWP (assembly) does not equal IWP (prefab).

The pipe fabrication priority is based on manufacturing efficiencies such as size piping and similar metallurgy. Structural fabrication is also based on size and type of members, rolled versus flanged versus plate girders. The module assembly is based on a set sequence of the main structure, secondary structure, equipment, piping, instrumentation, and electrical by level.

8.4 Selecting a Fabrication Yard

So, how do you decide which fabrication yard to use?

We did a quick review of this topic back in Chapter 5 in Step #3 under the Analyze Module Potential of the Business Case Model and talked about some guidelines for fabrication selection, suggesting that "size," "number of yards utilized," "location," and "complexity" all matter and are critical to proper selection. Here is a bit more on these subjects.

8.4.1 Project Size

The proposed job needs to be big enough to be noticed but not so big that it overwhelms the fab yard's capabilities. A rule of thumb heard is that the proposed job should target about 60% of the nominal capacity of the fabrication yard. A job using most of the fabricator's resources will command the attention and perhaps respect of that fab yard's management. Remember, they are in the business of completing tonnage throughput. If your job uses most of their resources, they will have more trouble starting or continuing other jobs and will want to move yours out, if for no other reason than to line up the next job. Also, if there are multiple jobs in the fab yard, the larger one should get the "better" supervision and perhaps craft leads, since any slip-ups could impact production, which impacts their ability to get the job out on time (and the next one in).

8.4.2 Number of Yards Utilized

For very large projects, considerations are necessary in terms of the option of using multiple yards.

- Advantages of using more than one yard:

  - Being able to maintain a yard capacity closer to that target of 60% to avoid overwhelming the single yard.

  - Using both smaller and larger yards or yards better suited for certain types of modules (pipe rack vs. equipment).

  - Having the flexibility to divide the module up by complexity or delivery sequence.

- Disadvantages include:

  - Requiring a separate construction management team (CMT) at each fab yard.

  - Additional complexity in procurement with materials split between several yards, plus the site.

8.4.3 Location of the Yard

A good visual of the global presence of many of the major fabrication yards is shown in Figure 8.2.

Even without the fabrication yards of South America, the Middle East, and some in Africa identified in Figure 8.2, it is obvious that a suitable fabrication yard can be found almost "anywhere in the world." As such, the issue in optimizing the selection is not just finding one or more that suit the project but finding ones that are "convenient" in terms of cost, support, and shipping. A couple of considerations in terms of location:

- **Construction Management Team (CMT).** Remember, each fabrication yard will require some sort of CMT. Obviously, the simpler pipe rack scope of work will require less supervision than the multi-disciplinary procurement, expediting, and quality assurance/quality control (QA/QC) requirements of heavy equipment modules. With a CMT ranging in size from a minimum of 6–10 people for a local fab yard to over 100 (including many ex-pats) for effective supervision in an international yard, this is an important decision.

- **Proximity to site or each other.** Consider a yard mix that will optimize current project personnel.

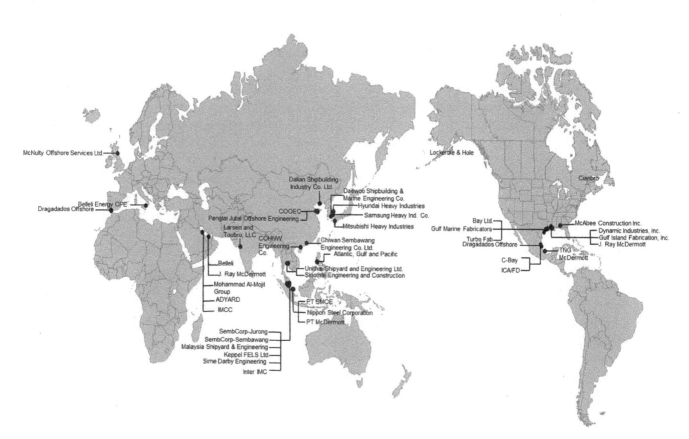

Figure 8.2 Locations of sample fabrication yards worldwide.

For example, two yards close together or a yard "close" to the project site may allow one CMT to monitor both.

8.4.4 Complexity

Some modules will be easier and faster to fabricate and assemble, requiring fewer and less-skilled workers. Pipe rack modules are relatively simple and straightforward, with mainly steel and piping and (perhaps) a few control valves, instrumentation, cable trays, and sometimes an aerial cooler (fin fan). But these same simple pipe racks typically take up a lot of real estate in terms of their footprint when being assembled.

On the other hand, equipment modules can be sophisticated, multi-disciplinary, multi-level structures that require a smaller fab yard footprint. Yards specializing in these are typically located in a more highly urbanized area, where land is less available, and costs are higher. These fab yards must get the most out of their infrastructure investment and are looking for craft workforce-intensive modules that occupy a smaller physical yard footprint.

So, the best fit for the pipe rack modules is fab yards with more assembly space and a workforce more suited to the assembly of structural steel and piping. Such yards are typically located more in the developing nations or outside of urban areas, where land is more available and craft is not as specialized. Conversely, the best fit for equipment modules is fab yards that have ample and skilled crafts available.

Why is this important? It does your project no good to try to fit a module type into a fab yard that is not suited (or not interested) in building that type of module. For example, the Far East fabrication yards (e.g., HHI, SHI, Bomesc, COOEC, etc.) not only have the skilled craft required but also have a business model and infrastructure that support the procurement and design necessary to carry out the highly complex module assembly effort. On the other hand, some of the fab yards in SE Asia and the Middle East have more assembly space as well as a labor force more suited to the assembly of the less complex pipe racks. Both groups will provide a better project value if they are contracted to work on modules complementing their strengths.

8.5 Contracting Strategy

Along with selecting the fabrication yard, there are considerations concerning the contracting strategy. These can be broken down into two main considerations.

8.5.1 Type of Contract

Purchase order versus subcontract, which is better?

We have seen modules get built under both, but in terms of which one works better, it is the *subcontract*. The subcontract provides the project with the ability to maintain greater oversight during the execution, which allows the project more involvement in terms of a management team (construction, QA/QC, safety, engineering, etc.). Since the equipment and pipe rack module fabrication and erection are more aligned with site construction practices due to the higher labor work-hours, this fits more appropriately with the typical subcontract documents and templates.

The use of a purchase order (PO) would not provide the robust terms and conditions found in a subcontract for erection activities. The PO would not allow you to have the management team present during the execution and sometimes could be limited to simply a shop inspector. While POs are designed for equipment fabrication, when there is a combination of steel erection, pipe erection, equipment erection, installing cable trays, etc., *the PO is not as efficient*.

In summary, the erection execution would be the biggest factor in deciding to go with a subcontract—in this case, the module erection is like construction work, only off-site (in effect, a mini-project).

8.5.2 The Best Type of Subcontract

While the "best" subcontract from the project standpoint for a fabrication effort is a *lump sum*, typically, there is never a complete enough design effort at the time the fabrication bid is sent out to avoid all the "extras" that will crop up during the actual fabrication, as the engineering is completed and sent to the fab yard.

What seems to work best is a contract that ties down parts of the fabrication effort that can be confirmed but leaves the variable other parts flexible to allow changes without impacting the validity of the entire contract. So, typically, the contract is constructed to set the fab yard's indirect costs (the ones that cover management, site use, office space, etc.) as a fixed cost based on the project size and details. Then, that contract sets the portion of the contract relating to the cost of materials as a direct pass-through with a percentage increase for administrative support and the actual cost of the module fabrication and assembly based on a unit rate per quantity produced.

For example, the structural steel would be priced based on a price per ton, based on size (light, medium, heavy, extra heavy). The piping would be priced by diameter inch (dia. in.) of weld required. There would be other metrics in terms of electrical and instrumentation, with the aim to quantify as much of the variable costs into unit rates that would then be multiplied by the number of units produced.

The fab yard would be required by contract to honor these unit rates regardless of the quantity change or growth within a certain variability called a "dead band" that is also set by the contract. This "dead band" would be calculated as a ratio of the actual as-built quantities over the originally bid quantities. Typically this should be set as large as the fab yard would allow (e.g., variation of +/- 10% to +/- 20%). As long as the scope growth did not exceed this dead band, the fab yard would still need to support the module fabrication effort with the originally bid management and other indirect costs. Should the dead band be exceeded, then the contract would be subject to re-negotiations (because the scope had grown so much that there would need to be an increase in the fab yard management team, personnel, etc., to cover this larger scope).

Besides the contracting method, we acknowledge that implementing the principles and methods of an integrated project delivery (IPD) can be an ideal project delivery approach for industrial modular projects. The CII Research Team-341 highlighted "Preassembly or Modular Construction" as one of the 20 implementation methods of IPD (Construction Industry Institute, 2019). There is no study investigating the effect of IPD on industrial modular projects yet. But we share the vision that if IPD is executed appropriately, modular projects can also benefit from it by increasing collaboration and integration.

8.6 Division of Responsibility

Along with the increase in complexity that some fab yards can accommodate comes an entire set of support services that these fab yards may also offer. These range from simple bulk materials procurement and expediting to detailed design, third party equipment purchase, design and fabrication of the specific equipment, and fabrication of structural members and process spool piping.

Because the larger fab yards have these material procurement and expediting services already built into their organization, the fab yard has a strong desire to utilize them. This makes it important that the project begins discussions on the division of responsibility between what the project will be designing, purchasing, and fabricating and what can be assigned to the fab yard to handle. Having the fab yard handle some or all of these scopes of work reduces interfaces between companies, which minimizes the potential for inefficiencies and change orders. The downside is that the owner/client and EPC must now follow up with the "extra" QA/QC efforts to ensure that the project is satisfied with the quality coming from these internal fab yard efforts. It also makes the owner/client and EPC rely on the fabrication yard to actually perform these activities in a timely manner.

So, early in the project, a division of responsibility document should be developed, which outlines what scope will be handled by whom. It can be as simple as the basic commodity chart shown in Figure 8.3 or complicated and involved with detailed specifics on virtually every item to be purchased or built. And, since fab yards differ in their abilities, this division of responsibility will be unique for each fab yard being considered. Certain items in Figure 8.3 have been highlighted as examples of items that might be changed in terms of responsibility based on the fabricator's abilities or the project's desires for closer control.

#	Commodity (excludes major equipment)	Purchase/Provide	
		EPC Contractor	Fab Yard
1	**Structural Steel**		
1.01	Structural Steel Shapes		X
1.02	Stairs, Handrail, Ladders, Cages, etc.		X
1.03	Steel Fabrication		X
1.04	Permanent Erection Bolts		X
1.05	Shims		X
1.06	Slide Plates		X
1.07	Touch up Galvanizing Material		X
1.08	Misc. Steel - Field Fab Material		X
1.09	Fireproofing		X
1.10	Seafastening & Grillage		X
1.11	Temporary Bracing		X
2	**Above-Ground Pipe**		
2.01	Fabrication (Large Bore and Small Bore)		X
2.02	Pipe Coatings		X
2.03	Pipe, Fittings, Flanges		X
2.04	Specialty Pipe, Fittings, Flanges		X
2.05	Bolts & Gaskets		X
2.06	Permanent Manual Valves	X	
2.07	Control Valves	X	
2.08	Cryogenic Valves	X	
2.09	Temporary / Testing Valves		X
2.10	Vents / Drains for Hydrotest		X
2.11	Temporary Erection Supports		X
2.12	Temporary Pup Pieces		X
2.13	Weld Rods and Consumables		X
3	**Piping Specialties**		
3.01	Strainers	X	
3.02	Safety Showers	X	
4	**Engineered Supports**		
4.01	Spring Hangers	X	
4.02	Snubbers	X	
4.03	Cold Shoes	X	
5	**Standard Supports (Field Fab),touch up & testing**		
5.01	Touch-up Paint		X
5.02	Thread Lubricants for Flange Bolts		X
5.03	Tape and Thread Sealants for Threaded and Coupled Pipe		X
5.04	Initial End Cap Protection for Pipe and Valves		X
5.05	Hydrotest Water		X
6	**Above-Ground Electrical**		
6.01	Breakers / Switchgear	X	
6.02	Transformers	X	
6.03	Substation/Powerhouse	X	
6.04	Motor Control Centers and Switchgears	X	
6.05	Cable / Wire	X	
6.06	Cable Tray / Raceway / Hardware		X
6.07	Conduit and Hardware		X
6.08	Electrical Panels	X	
6.09	Receptacle and Switches	X	
6.10	Termination Kits		X
6.11	Splice Kits		X

#	Commodity (excludes major equipment)	Purchase/Provide	
		EPC Contractor	Fab Yard
6.12	Lighting / Power Transformers	X	
6.14	Light Fixtures	X	
6.15	Flood Lights	X	
6.16	JBs / Bulks	X	
6.17	Support Stands		X
6.18	Supports (Cable Tray and Conduit)		X
6.19	Electric Tracing System		X
6.20	A/G Grounding Connections		X
6.21	Telecommunications	X	
6.22	Touch-up Paint		X
7.0	**Instrumentation**		
7.01	DCS	X	
7.02	Instruments	X	
7.03	Analyzers	X	
7.04	CVs	X	
7.05	RVs	X	
7.06	Transmitters	X	
7.07	Vortex Meters	X	
7.08	Level Gauges	X	
7.09	Pressure Gauges	X	
7.10	Temperature Indicators	X	
7.11	Thermo Couples	X	
7.12	Orifice Plates	X	
7.13	Fire & Gas Detection / Alarm Systems	X	
7.14	Cable / Wire	X	
7.15	Cable Tray / Raceway / Hardware		X
7.16	Instrument Panels	X	
7.17	JBs / Bulks / Tubing		X
7.18	Panel and Instrument Support Stands		X
7.19	Touch-up Paint		X
8	**Insulation**		
8.01	Cold		X
8.02	Hot		X
8.03	Acoustical		X
9	**Equipment**		
9.01	Process	X	
9.02	Utility	X	
9.03	Bolts		X
9.04	Permanent Gaskets		X
9.05	Temporary Gaskets		X
9.06	Touch-up Paint		X
9.07	Shims		X
9.08	Level Plates		X
10	**Loadout**		X
11	**Center of Gravity**		X
12	**Dimensional Control**		X

Legend

▓ : potential scope kept by EPC contractor

░ : potential scope transferred to Module fabricator

Figure 8.3 Module commodity scope split.

8.7 Summary

In this chapter, we went over some basics about the "generic" fabrication yard, including the benefits and characteristics, selecting the right fabrication yard, contracting strategy, and division of responsibility.

Reference

Construction Industry Institute (2019) FR-341 – Integrated Project Delivery for Industrial Projects. Austin, TX: The University of Texas at Austin: Construction Industry Institute. Available at: https://www.construction-institute.org/resources/knowledgebase/knowledge-areas/project-program-management/topics/rt-341/pubs/fr-341.

chapter 9

Module Considerations by Project Group

This chapter will examine some of the finer points that can make a good module project great or, if not followed, can really make it painful. Many of the examples these points are based on are from large module jobs, either in terms of actual size or number, in some cases, with complex process designs. There is nothing unique about these big, complex module jobs or their challenges. It is just that when there is a failure to follow good module planning on such large projects, the consequences can be dramatic and memorable.

Another way to look at the differences between the traditional stick-built approach and the modular execution philosophy is by the major project group. There are subtle (and not so subtle) differences that must be adhered to in order to make the project function smoothly in the early planning phases. And, just as there was a continued development of project details in the 13-step module feasibility flowchart in Chapter 5, there is also an evolution in the detailed approach by each of these groups as the project progresses, with initial guidance philosophy established early, followed by filling in the details to promote the timely accomplishment of the various tasks.

We will explain each of the different module considerations in the project in terms of how the module aspect impacts them. The subjects included in the module considerations are:

- Engineering
- Schedule
- Procurement
- Sub-Contract
- Fabrication
- Completion/Testing/Prep
- Load-out
- Module Movement
- Shipping,
- Construction.

We will walk through these considerations in the order they might typically surface on a module project. This will be a lot like the discussion on the Execution Plan Differences, but hopefully, illustrated with actual examples and details that will make more sense of the explanation in Chapter 6.

9.1 Engineering Considerations

We start with Engineering due to all the big and little things that can create bumps along the road if not addressed at the right time.

9.1.1 Module Evolution

The "module" does not just spring out of the minds of the engineering group. While it can begin with a detailed 3D model approach, it typically starts out as a simpler sketch or idea. The transition from the stick-built layout to a modular layout typically means moving equipment around, especially in the vertical third dimension. So, whatever the development method, it needs to be versatile and easy to adjust as the equipment moves.

The sketch or idea can be a two-dimensional sketch or an electronic version of the same drawing. Figure 9.1 shows how one particular module was developed. Starting with the major equipment proposed for the module, the equipment is laid out in terms of basic elevation and orientation (see Figure 9.1). Along with this effort, the next was some very early considerations regarding how the modules would be situated on a particular heavy haul transport vessel (Figure 9.2).

9.1.1.1 Path of Construction

Along with this early effort of module equipment development is the "path of construction" effort that is evolving at the project site. In its earliest considerations, this path of construction becomes an attempt to determine how to get this module, along with all the others laid out on the plot plan, and develop a logical method to move them onto the site and into their final positions. The considerations include ensuring that the sequencing will allow other site activities to progress to the point that the underground work would be completed before the module arrives. In fact, this is one of the benefits of modularization—it can

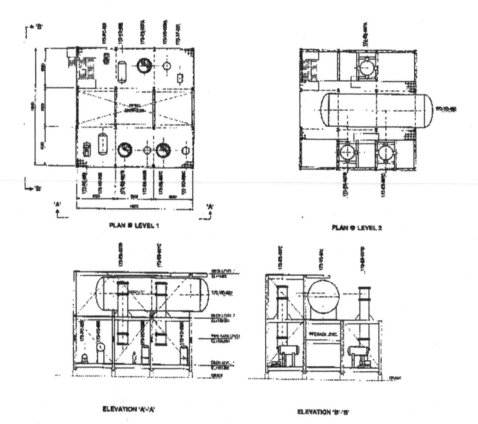

Figure 9.1 Module preliminary equipment layout.

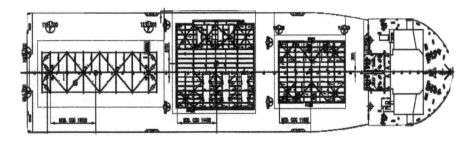

Figure 9.2 Preliminary module layout on RO-RO (Roll-On-Roll-Off) vessel.

provide additional time to perform the underground installations at the site. In addition, a well-laid-out plot plan would provide alternative "paths of construction," giving the project more potential flexibility in their work activities should something unforeseen happen, like a delay of major equipment, site slowdown, labor issues, etc.

As the module design progresses, the simple two-dimensional module becomes a 3D model, using any one of several commercial software programs. Because these are exact representations of the equipment, piping, and steel, there is a tremendous amount of engineering information that must be developed to be able to properly and accurately incorporate all the equipment, piping, steel, and electrical and instrumentation and controls (E&I&C) into the model.

The great benefit of the 3D model is that you have an exact "picture" of the module that, because of the information it contains, you can actually build from it. The downside is that you must have all the engineering data to build the pieces and model accurately. Anything less and the model will not be useful in terms of a fully designed effort. This puts a burden of expectation on all the groups supporting the model development, from the early module layout piping designer (who must have the correct dimensional data on equipment as well as an owner-approved complete catalog of electronic piping components) to the structural engineer (who must have weight and size information on everything in the module to properly size and locate all primary, secondary, and eventually tertiary structural steel) as well as overall weight information for a proper design of the foundations the module will rest on.

Besides the obvious benefit of having an exact representation of the final plant, some of the more subtle benefits of the 3D model are in the analysis of operations and maintenance (O&M) access and clash checking.

So, this same early module sketch shown in Figure 9.1 now can be developed in a complete 3D model, with, for example, dimensional piping data that can be taken directly from the model for use in material take-offs as well as piping isos which will eventually be fabricated into piping spools by a pipe shop or fabrication yard (Figure 9.3).

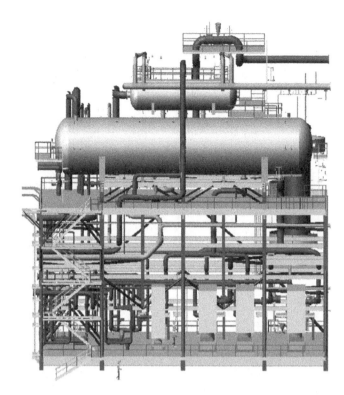

Figure 9.3 3D module model development. Source: Courtesy of KBR.

9.1.2 Operations and Maintenance Input

O&M experts are very knowledgeable on how the plant needs to operate and how they need to access the various parts of the equipment. They can provide early identification of accessibility requirements as well as confirmation of the adequacy of paths already provided. By participating in O&M "virtual walks" through the model, the O&M experts can quickly identify any difficult or near impossible access to piping or equipment due to poorly placed structural steel and piping. To maintain the major equipment, human access routes should be identified. The routes are laid out in the 3D model as virtual "solid" volumes in the module that cannot be inadvertently penetrated by piping or steel.

These O&M reviews are critical and should be scheduled as soon as there is a preliminary module layout. Failure to identify and address the operator's concerns early enough typically results in re-work of the module itself. At the very least, it results in a very unhappy operator. On the positive side, these early "model reviews" can provide opportunities to consolidate, reduce costs, and optimize equipment layout by understanding exactly what

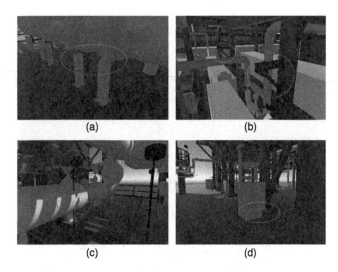

Figure 9.4 Examples of detecting design errors with 3D model. Source: Han and Leite (2021). Reproduced with permission of American Society of Civil Engineers (ASCE).

is required from an O&M standpoint and providing extra considerations and accommodations only where they are necessary.

A good example is the optimization of the ladder access to the top of all tall vessels. Instead of adding expensive access on the exterior of every tower, the design process is optimized to only add access where required by O&M via interactive discussions. Rarely required access on the rest could be scheduled with other crane-required activities, reducing both initial capital costs as well as yearly maintenance costs.

The model reviews also typically incorporate "clash checking," where the 3D model provides a listing of the places where the design has resulted in two or more "solid" objects (e.g., pipe & pipe/pipe & steel) occupying the same coordinates in the model, and in effect run through each other. See Figure 9.4 for examples of detecting design errors with 3D model from (Han and Leite, 2021). Timing and duration of these model reviews and clash checks depend on the module maturity, level of accuracy in the initial inputs, and complexity of the module, but typically are scheduled on a regular basis to avoid too much re-work.

So, why all this emphasis on the engineering design development? As previously mentioned, these modules typically end up on the project's critical path. As such, any delays in the 3D model development of each module

could impact the fabrication, assembly, transportation, or setting and need to be eliminated. The 3D model of the module itself becomes a "mini-project" or a "project within the Project."

9.1.3 Structural Engineering Is King

Essentially, everything on the module, from the design to the structural steel, equipment, piping, and all other materials, must meet the fabricator's schedule or be accelerated. At the very bottom of all this (literally) is the need to get the structural design correct. This means that all the details from equipment to piping, to operating pipe stresses, to supports for shipping and transport must be developed at least on a macro scale in time to complete the structural design and make the correct bulk steel purchase. This also means that the overall dynamics of the engineering disciplines must change. With everything on the module being supported by the structural steel, and the structural steel purchase and fabrication being the first activity of the module fabrication yard, the entire module design is driven by the early engineering of the structural group. It is this group that dictates when information must be provided in order to have enough time to develop the design to the point needed to provide the necessary structural material take-offs (MTOs) to the fabrication yard so the steel can be ordered from the steel mills in time for a delivery and start of the structural steel fabrication.

This becomes a bit of an issue when working with the traditional project engineering hierarchy. For the typical stick-built project, the piping design discipline is the critical path, and, as such, they provide much of the input to the project timing needed to complete their design efforts. As noted above, with the module project, the structural steel design becomes a more critical need and, therefore, a higher priority. As such, the structural design group must drive much of the timing of the engineering discipline approach, and this drives the equipment design, procurement, and delivery of early engineering details.

For the traditional engineering design team, this can be a bit of a transition problem and must be monitored carefully to ensure that early on, the structural design team takes this responsibility and is not relegated to the back of the engineering discipline pack when it comes to being provided their design needs. An engineering team that has previously worked on a module project will understand this slightly different emphasis and make the early engineering efforts much more seamless.

With respect to coordinating with the fabrication yard, this is an important aspect of developing an efficient structural design. Early decisions on the use of plate girders and tubular columns as well as welded versus bolted structures impact the basic design and must be made at the outset of the structural design. We will cover this in a bit more detail in Section 9.5 on Fabrication Considerations.

9.2 Scheduling Considerations

One of the earliest module efforts is the development of a project schedule. This is not the original level 1 project schedule that shows the big picture items flowing from engineering, to procurement, construction, and start-up. The schedule of interest is the subsequent ones that begin to look at the module concept in terms of the modules being a set of smaller EPC projects within the overall project. This version of the project schedule should begin to look at multiple small projects with interrelated completion dates to get each completed module on its transportation vessel and to the site in the correct sequence for installation. This is where the scheduling gets a bit more interesting.

Each module is composed of numerous pieces of equipment, each with its own ties back into the engineering design effort, which goes even further back into the project process concept phase, where equipment is being identified. One of the early considerations in the selection of one particular process over another in the assessment phase is the equipment itself, its overall complexity, and all the issues with the development of it. Of course, the ease of equipment development with respect to timing and module construction will never drive the process selection, but when going through the assessment process, the timing of such equipment delivery should be considered in terms of how the equipment will interact with the module that it will be going into.

It's really not much different than the stick-built schedule approach where specific equipment, labeled "critical," is identified early in terms of delivery to the site to determine what and how much "early" engineering and procurement effort must be made typically prior to the financial investment decision (FID), in order to ensure that this equipment will not hold up the construction phase of the project. However, with the modular project, this equipment may be needed much earlier in order to be installed in a module as opposed to having the luxury of being able to wait until required at site dates.

Why is this differentiation so important? There will be pieces of equipment that, for one reason or the other, just cannot be accelerated in the schedule to meet the module assembly date. These pieces of equipment need to be given extra consideration in terms of how to efficiently develop the module plan (e.g., with or without them in a module) and what omitting them from the module will do to efficiencies and ability to move work-hours off the site.

For example, it may be that the simplicity of hook-up on a very large tower does not warrant the extra efforts and extra module steel to support it internally in the module, and setting later at the site on an adjacent concrete foundation makes the most sense. The same thing could be said of large rotating equipment that produces vibration and harmonics that are difficult to design for in the module. The best solution for these types of equipment may be to design to set adjacent to the supporting module and deliver directly to the site, providing more time to build that piece of equipment. These are the easier decisions to make.

The more difficult situations are where a large number of work-hours are associated with this equipment hook-up at the site. Examples include the motor control center (MCC) or remote instrument unit (RTU) that requires extensive hook-up work-hours to other equipment within the module, making it desirable to pre-install it in the module at the fab yard. However, completion of these complex units is almost always an issue because the E&I&C disciplines are the last engineering groups that complete their designs, in part due to the inputs needed in terms of motor and pump sizing, which depends on the completed design of all the sub-systems.

So, early discussions with the module fabrication are strongly suggested to determine how a module can be designed to allow for the late installation of such equipment. This is why it is so important that the module definition be determined very early in the 13-step module flowchart of Chapter 5.

9.3 Procurement Considerations

After the project has worked with scheduling to develop the plausible "path of the project" in terms of the selected number of modules, what they will look like and equipment contained in them, and the probable sequencing of their shipment and installation, plans need to be developed on how to obtain all the specific engineering details in terms of equipment size, support, nozzles, and potential piping configuration to allow the structural team to develop the module steel design.

This is where the procurement contracting strategy becomes critical. Details on how the project will get approval from the owner/client in terms of equipment purchasing will dictate how the procurement group will approach each equipment vendor. Strict requirements for the typical "3 bids and a buy" might result in the need to push the engineering of this equipment earlier (to the left) so that this bidding activity can be started.

If the owner/client cannot approve an early purchase (as in pre-FID), options need to be discussed regarding the commitment of some pre-FID funding to perform the required bidding activity along with the conditional award to the successful bidder. Such contracts stipulate that the project is committing to the vendor that they will be awarded the contract IF the project is approved. But, in the meantime, the vendor is contractually obligated to start detailed design to meet schedule, understanding that if the project does not get funding, they will only be paid for the early engineering efforts defined. This will enable the completion of the necessary equipment design details to hand over to the structural group for the module steel design.

The owner/client must commit to enough pre-FID funding to reimburse all these critical equipment vendors for this early engineering. If pre-FID funding is not available (as with some Middle East projects), an analysis of the impact of this late design information on the module development must be made as it could impact size, content, and number of modules the project will be able to "fit" into the owner/client schedule.

9.3.1 Fabrication Yard Interfacing

Working with one or more fabrication yards is also an option to alleviate late equipment deliveries. The timing of this interaction should be concurrent with the start of the module design, as the type of module designed should complement the strengths of the fab yard's capabilities in terms of assembly. At this same time, discussions should also be initiated with respect to possible options for late incorporation of delayed equipment to see if work-a-rounds are available.

Remember, if the goal in a module project is to move work-hours from the construction site to somewhere where it is cheaper, the project needs to consider EVERYTHING that can provide the opportunity to maximize this goal.

9.4 Sub-Contract Considerations

With the need to get fabrication yards involved early, it is important that the project puts in place an effective and efficient sub-contracting organization that can begin working on many of the potential sub-contract needs early.

As mentioned earlier in this chapter, the sub-contract document provides more robust contract terms and conditions for involving the fabrication yards and the other main players in the module project than the purchase order. This is because the sub-contract can provide the project management team with a greater oversight capability in terms of involvement (construction, QA/QC, safety, engineering). Other activities that should be considered in sub-contracting beyond the typical sub-contracts for a stick-built effort are as follows.

9.4.1 Heavy Haul (HH)

Typically involving a third party team from international transport companies, such as Mammoet or Fagioli, who can supply crews and equipment needed to weigh, move the modules to the quayside, load on transport vessels (tie-down and sea-fastening typically provided by the fab yard), offload, transport to location, and set up at the project site.

9.4.2 Vessel Transport

Typically involving one or more large roll-on-roll-off (RO-RO) or self-geared vessels that have the crane capacity to self-load. In either case, these companies perform various stability analyses as well as develop (or confirm) the grillage and sea-fastening design for attaching or tying down these modules to the ship's deck.

9.4.3 Module Offloading Facility (MOF)

The module offloading facility (MOF) is much more than just the typical offloading dock for construction bulks. It will require additional design considerations in terms of size and capacity as well as how it will interact with the rest of the shipping activity (e.g., bulk materials, general cargo, and any proposed product shipments).

9.4.4 Heavy Lift (HL)

This sub-contract will need to be adjusted in terms of how the typical stick-built heavy lift equipment (e.g., cranes, cherry pickers, boom lifts) interact with the HH contractor bringing in the modules. The goal would be to minimize overlap without creating gaps in the coverage. Special tower cranes may need to be added in the proximity of the modules to assist in final hook-up needs.

9.4.5 Pipe Spools and Galvanized Structural Steel Fabrication

For the smaller module assembly yard that does not have either pipe spooling or structural steel fabrication capabilities, these two activities can end up also being sub-contracted. Both require a certain amount of inspection (dimensional and the quality of galvanized coating) and the interface in terms of fabrication priorities and shipping to the module yard, making them more fit for sub-contracting.

Early coordination of all of the above is critical to avoid gaps or overlaps in terms of responsibilities. For example, the fab yard typically owns some self-propelled modular transporters (SPMTs) or multi-axle trailers and may require the HH company to utilize these. With respect to module weight, the HH contractor has the ability to weigh the modules via the SPMT hydraulics. However, the need for accurate weight and the center of gravity is often required much earlier than final load-out as it provides early input to the vessel contractor for some of their stability analysis. This may mean bringing in a separate third party to weigh the modules.

In particular, coordination of other third party materials and fabrication support during the engineering effort must be done way before the start of fabrication to ensure the engineering team's working priorities are consistent with the module fabricator.

9.5 Fabrication Considerations

One of the more critical decisions to be made very early in the module design has to do with the module structure itself—the type of steel members and their protective coating.

9.5.1 Bolted vs. Welded

Historically, the break between bolted structures and welded structures was related to the *overall size of the module*, with the larger modules being welded. Part of this had to do with the history of the module fab yard. Historically, the larger offshore decks were built in the Far East, where the use of welded plate girders was required due to the large size and weight of the offshore structures. In addition, tubulars were used for the major columns. These provide increased strength in all directions, and the tubular connecting points, while complicated to miter, were relatively easy to weld. The fab yards that built these offshore deck structures were very familiar with the tubular connections from the huge offshore jackets they also built.

As they evolved to more onshore modules, the higher efficiency of the tubular sections for vertical members and the large custom-made plate girders also made sense for these land modules. So, the Far East, Korean, and SE Asian yards continued with the welded structure (which dictated that they be blasted and coated with a multi-coat combination—typically of inorganic zinc and epoxy). The fab yards had extensive experience and a vast number of qualified welders to support this fabrication method.

In the US and along the United States Gulf Coast (USGC), the module fabrication business (with a couple of notable large fab yard exceptions) grew as an evolution from the stick-built project philosophy. This was based on pre-fabbed structural steel that was designed, fabricated, and galvanized to be bolted together, very similar to what had been done in the stick-built projects. The only difference was this effort was being handled offsite in a fabrication yard.

More recently, the differences have been blurred. Designers have developed very large (e.g., 1500 metric ton) galvanized and bolted modules that the Far East fab yards have efficiently assembled. Other small yards have become efficient in welding. The cheap labor drove some yards that used welded connections (that were more cost-efficient) to work with bolted structures requiring a less skilled workforce.

9.5.1.1 Some Fabrication Misconceptions

Bolted and galvanized structures have always been the favorite of the USGC stick-built projects. They are easy to put up, and the galvanized coating can resist corrosion for 20+ years. But with the transition to a modular project approach, the welded vs. bolted options deserve closer scrutiny. A couple of misconceptions follow:

- *A bolted structure is lighter—false*. The structural members are typically the same size, and the bolted connections more than offset the extra weld materials of a welded structure.

- *A bolted structure is cheaper—it depends on the construction crew skill sets*. In the large Far East fab yards where welding is historically king and labor is cheap, the welded module is very competitive and, in some cases, cheaper than the bolted structure due to all the handling required for galvanizing. For the smaller assembly yards where welding expertise is not as common or plentiful, the bolted structure is favored, with some fab yards not even able to quote a welded module.

- *Bolted is easier and faster to assemble—it depends*. The bolted structure requires much more laydown yard space. With a welded structure, typically, the structural members are cut to length with simple beveled or coped ends. The storage space required is minimal. With the bolted structure, the member fabrication is much more intensive, with welded clips, drilled plates, and all the interface issues of matching all the holes. Logistics is more complex than for a typical piece of steel. The complex fabricated piece must be packed up and shipped to the offsite galvanizing facility, then to the module assembly yard, where it is unpacked, sorted, and staged before finally being installed. All this handling increases time and cost, which reduces any benefit of cheaper assembly labor.

Figure 9.5 Welded pancake deck construction.

Figure 9.6 Completed welded pancake deck structure lift and set.

- *A bolted assembly is easier—it depends*. With a welded module, there are more options with respect to assembly. The completed deck level (or pancake) can be designed to be lifted and set on top of the lower completed module, as shown in Figures 9.5 and 9.6.

Bolted modules tend not to have this rigidity in only two dimensions, requiring they be assembled in place to complete a three-dimensional structural frame before moving, requiring more work at heights. Also, in some cases, the bolted structure requires more temporary internal bracing that limits through pipe access, as shown in Figure 9.7 versus a similarly designed welded structure of Figure 9.8.

Figure 9.7 Required internal bracing on bolted structure.

Figure 9.8 Open access on welded structure.

9.5.1.2 What Difference Does It Make?

- One obvious driver is **module size.** Members can only get so big before they become an issue with respect to the bolted connection size as well as the ability to fit them into a galvanizing vat. In addition, the large stresses potentially placed on the module while under construction and during movement make the welded connection stiffer and more resistant to slight differential movements that one might see with the bolted connection.

- The second driver is **the workforce.** A bolted and galvanized module is not much more than a large Erector set and can be assembled by a less skilled workforce than the welded module. This provides more flexibility when evaluating the different fab yards and less reliance on the welding quality of the crew.

- The third driver is **the owner/client**. A bolted and galvanized module is more similar to the existing plants in some areas like the USGC. The owner/client likes the idea of a structural member with a coating that is good for 20 or more years with no maintenance. Trying to sell them on a welded and "painted" structure, requiring maintenance within 10 years, is difficult, if not impossible.

At the end of the day, the combination of module size/client wishes/fab yard expertise must be considered when deciding on whether to weld and paint or bolt and galvanize the project modules.

9.5.2 Structural Shapes

A bit more on the nuances of best structural shapes. As mentioned above, large plate girders (in some cases up to 1.3 meters in depth) and column tubulars (in lieu of standard I or W sections) are used for the larger modules. Modules get larger as the equipment inside them increases in size, and it takes more square footage to be able to get a complete sub-system within module limits. This creates a need for larger access routes and maintenance routes to be able to get the equipment in and out of the module. Since you cannot always locate this equipment on the outside edge of a module, larger access ways become a requirement.

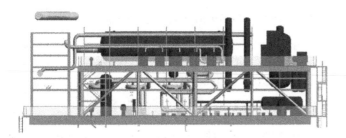

Figure 9.9 Tubular columns and bracing. Source: Courtesy of KBR.

The solution: the tubular column and tubular bracing. With tubular columns and bracing, you can create longer and wider spans between these supports, opening up more vertical and horizontal access (see Figure 9.9). Contrast this with the traditional W-shape that must be oriented carefully to make sure its weaker axis is not overstressed. This becomes even more onerous where large bolting flanges must now be included for the module designed for bolting.

9.5.2.1 Weight Control and Center of Gravity (C of G)

Critical to the overall coordination of shipping and module movement, as well as the design of the grillage and sea-fastening, is the weight control program. This effort should be kicked off as soon as the modules have been preliminarily sized and equipment within each one identified. With the advent of 3D modeling programs, this information is eventually incorporated into the model. However, during the early phases of the project, decisions on weight and size must be made in terms of information available since not all the detailed designs have been completed.

During the early phases of design, there are numerous ways to develop an approximate weight and an approximate C of G. For the typical pipe rack, it is fairly uniform in density, and the C of G is not too far from the geometric center of the pipe rack volumetric shape. Structural steel weight can be assumed from a similarly sized, previously built, stick-built pipe rack. Piping weight can be assumed from an average pipe size times the estimated total length of piping within the pipe rack.

The equipment module can be a bit more complicated. Again, piping can be assumed to be more or less uniform throughout the module by level and based on a similar

average pipe size by total length estimated for that equipment module. Structural weight can be initially estimated based on some typical beam sizes or from a historic module of a similar size previously built and also uniformly distributed. Early weight estimates for the major equipment are then placed where each of their individual C of G's hits.

Depending on the accuracy of these estimates, a sliding contingency scale is added to piping, structural, and equipment. Contingency typically can start at 20–25% for the first pass during concept screening and is reduced in a stepwise manner based on the level of design development all the way down to about 3% for the completed design.

The goal in any module weight tracking program is shown in the lift weight trend graph (Figure 9.10). The goal is to keep the top tracking module estimated weight line plus contingency weight always above, but eventually converging to the bottom weight line (actual weight of all components) but never crossing it. As such, by using the contingency weight for all your stability and transport analyses, you will always be on the high or safe side.

Figure 9.11 is a typical load cell used in the actual module weighing when complete. The individual cells are placed under the module in enough points and hydraulically pressured to collectively "raise" the module off its fabrication supports. They are interconnected to provide a composite weight read-out. This module weight "final report" should be pretty close to any quantity survey results being tracked in the fab yard.

9.6 Completion/Testing/Prep Considerations

Just as in any project, the last 2% or 3% of project effort seems to take forever, and most projects do not get to 100% without a lot of extra effort. The same is true for a poorly planned module job, except that in the case of the incomplete module, the ramifications are far worse and more financially painful.

Why? Because the module project schedule is set up to deliver a module that is 100% complete. There typically are no allowances for shipping late or shipping

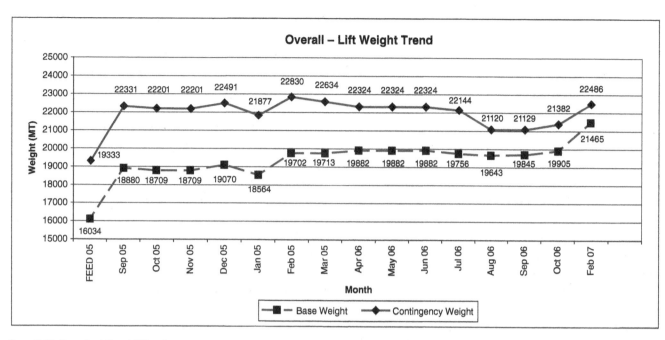

Figure 9.10 Example of lift weight trend.

Figure 9.11 Weight cell set-up used in module weighing.

incomplete. The project site does not have the capacity to handle the extra work. The site typically does not have the bulk materials to complete it either. Remember, the original reason for modularization was to try to maximize the work-hours that were moved off the project site! On-site work is much more expensive, harder to staff with quality craft, and as a result, will be more difficult to complete in the short amount of time left in the project schedule. Module carryover work only means extra site work and overtime.

So, it is important that the fabrication yard has a solid plan for this completion effort and that all the details regarding testing, (hydro or other non-destructive examination. NDE) pre-commissioning (instrumentation and continuity checks, motor bumping, etc.), and final checkout and walk-through have been agreed to and documented well before the first module is even being considered for shipment.

What does this effort involve? Typically, it is one or more early meetings with the owner/client, the fabrication yard management, the project site team, project engineering, and project management. With the original project goal of maximizing the number of site craft work-hours moved to the fab yard, it is important that some of these hours are not transferred *back* to the project site through poor planning, poor coordination, and most importantly, poor communication.

How can this happen?

Take the simple example of hydrotesting and cleaning. Hydrotesting is where water is used to fill the pipe and a pressure pump is used to pressure up the water to a pressure desired. The ideal project situation is where the piping is tested at the fab yard, and the only inspection at the site is the NDE of the "closure" weld between two sets of already hydrotested piping, confirming that the weld is sound. What could be simpler?

Unfortunately, there are too many instances of just the opposite happening. An incomplete plan for fab yard hydrotesting is floated out to an incomplete mix of the required stakeholders needed for approval of the process (perhaps even with management missing). As a result, the paperwork for the fab yard hydrotesting is either incomplete, not signed, or signed by personnel who are not recognized by the project site management. Because the "proper" authorities have not signed off on the testing, the resulting test documents from the fab yard team are not recognized at the site as something that the project could stand behind and support. As a result, it is decided to re-test all the piping again at the site.

So, the project pays twice—once at the fab yard for testing, and, again, at the site to perform the same testing, but this time using extra craft paid at the premium costs of the at-site rates, as well as including a second round of cleaning and dewatering required to make the piping ready for process fluids.

A similar situation can occur with the closure weld philosophy mentioned above. Instead of getting all parties to agree that two pieces of already tested piping can simply be welded together with some non-destructive examination, such as x-ray, gamma-ray, or ultrasonics, the project will start second-guessing the quality and proper sign-off of the work and decide that it should all be re-tested and re-cleaned at the site, wasting a tremendous amount of time and money.

How do you get around this dilemma? The entire module completion/testing/and prep for shipping procedures should be agreed to by a project team consisting of BOTH:

• the technical representation of engineering and quality that can put their hands on their hearts and say with confidence that the proposed completion testing efforts will be adequate;

• and the project management from BOTH sides (EPC contractor and owner/client) who have the authority to collectively change the pre-commissioning and commissioning criteria to permit such work in the fab yard.

With the more complex modules becoming almost entire small sub-system units, complete with power, instrumentation, and controls, the project should entertain creative ways to move beyond the typical pre-commissioning efforts of hydrotesting and moisture removal. Options for the actual introduction of process fluids into the modular units should be considered where substantial gains in completion can be achieved by doing so.

If, for example, a completed LNG liquefaction plant is required, and it cannot be assembled and pre-commissioned on the site, it may be worthwhile to bring gas into the fab yard (in terms of LNG, a gasification unit) to produce gas that can then be run through the new LNG facility to confirm its ability to produce LNG before it gets to the site. (Sort of like test driving a car you plan to purchase.)

9.6.1 Shipping Incomplete

This will happen to you on your job. Maybe not on the first or second module project, but somewhere along the line, you will be faced with the decision to ship a module that has not been completed. This may have happened because of poor planning, but it might also happen because of an error in scheduling, a labor crisis, a couple of weeks of bad weather, etc. In any case, this part of the Module Perfect Storm will be yours to contend with.

So, at what point do you bite the bullet and ship incomplete? The decision will depend on a number of factors:

• What are the availability and flexibility of your module transportation?

  • ocean-going vessel booked for the next several months?

  • readily available truck and trailer?

• Can the project schedule stand to have a module wait until the next vessel is available, or is it better to double up with craft labor at the site to complete the module?

• What makes more sense from a cost standpoint?

  • Pay demurrage on the ocean-going vessel to stand by and wait for the module to be completed?

  • Ship without a full cargo?

- Pay penalties to release the vessel from the contract and pick them up later?

- How is the late module going to impact the site "path of construction?"

  - Can the on-site team absorb the extra scope?

  - Will the extra work become the new project critical path?

Surprisingly, this should not be an issue that suddenly surfaces. Warning signals should have been everywhere and visible if someone on the team was paying attention. For example, a trend of ever-increasing missed intermediate completion dates and goals fails to get noticed. Or, craft productivity or even actual craft numbers have declined below scheduled requirements, but nothing is done about it.

Obviously, as soon as identified, discussions on alternatives to get the project back on schedule should be initiated. In addition, assuming this may not be enough, further options should be developed on how to accelerate the current completion efforts ahead of the current project schedule, understanding that whatever remedial efforts are initiated, they will probably not be as successful as originally sold by the team. Then, the agreed plan as well as the acceleration alternatives should be contractually binding.

9.6.2 Ship Loose

Even with all the efforts above, there will still come a time in your module career where there is nothing more you can do (or, more accurately, there is no more time for you to do it!), and you must prepare for shipping incomplete.

A couple of thoughts on how to ease the pain at the project site where the team is receiving this incomplete module:

- Provide a detailed listing of what is still incomplete. Discuss it with the site project management and send it to them well ahead of the incomplete module.

- Provide all the materials, including extra lengths of pipe (for each size and grade to be worked on), bolts, gaskets, and any other materials needed to complete the work. Do not assume they have them at the site.

- Organize the materials and work into work packs that are cataloged.

Figure 9.12 Ship loose items lashed to the deck (1).

Figure 9.13 Ship loose items lashed to the deck (2).

- Place pipe spools in the area(s) they will be installed and lash them down (Figures 9.12 and 9.13).

- Consider using connex boxes for organization and packing, shipping these along with the incomplete modules (Figure 9.14).

- Assign an individual to this effort, whose sole responsibility is to ensure all the above is handled properly.

9.7 Load-out Considerations

Load-out and *shipping* go hand in hand, but we will address them in the order that they actually happen, understanding that shipping design is addressed much

Figure 9.14 Ship loose items organized in connex boxes.

Figure 9.16 Module moved with SPMTs (1).

Figure 9.15 Trailer truckable module being loaded at indoor fabrication facility.

Figure 9.17 Module moved with SPMTs (2).

earlier in the project planning since these trip accelera-tion forces must be incorporated into the early stages of modular structural steel design.

9.7.1 Land Load-out and Self-Propelled Modular Transporters (SPMTs)

Load-out for transport across the land via roads is a relatively straightforward process, utilizing cranes in most cases to lift the module and set it onto the trailer (Figure 9.15). Where the module is extremely large or heavy, the module is typically built on short vertical sup-ports to make accommodations for the self-propelled modular transporters (SPMTs), a type of high capacity (large and heavy loads), highly maneuverable (360-degree

pivot) self-propelled flat top vehicle, that simply drives under the module and hydraulically lifts its deck to pick the module off of these construction supports. These SPMTs are 4-axle or 6-axle units that can be mechani-cally connected together end-to-end to form individual "lines" of the required length.

Figure 9.16 is a 14-axle line made up of 2–4 axle units and 1–6 axle units mechanically connected. These "lines" can then be connected side by side to form an any size length and width, integrated, moveable plat-form required to support whatever size module needs to be moved. The entire set of integrated "lines" is maneu-vered by a joystick-type computer console connected by cables to the SPMT power packs, which provides both locomotive and hydraulic power to the wheels. The console is typically strapped to an operator who walks alongside since the SPMT typically moves at walking speed (Figure 9.17). With the SPMT lines connected side to side, the multiple SPMTs can be operated as a single unit to transport wide loads (see Figure 9.18). Upon arrival at the site, the module is "set" on its foundation piers by driving between the piers and lowering the SPMT decks hydraulically.

9.7.2 Marine Onloading and Offloading

Offloading can be as simple as pulling a barge up on a shallow spot in the wider part of a bayou or river and rolling the module across some timbers (Figure 9.19) to the much more detailed and massive efforts involved in the onloading of a very large equipment module across

Figure 9.20 Onloading from dedicated module offloading facility (MOF) onto RO-RO vessel.

Figure 9.18 A module moved with multiple lines of SPMTs.

Figure 9.19 Offloading barge-mounted module rolling onto earthen ramp and timbers.

a module offloading facility, using multiple interlocking ramps and onto a specially built roll-on-roll-off (RO-RO) ocean-going vessel (Figure 9.20).

Onloading, transport, and offloading can also be handled by what is referred to as a "self-geared vessel," the terminology coming from the lift cranes that have been incorporated into the ship's hull. Such a vessel is capable of taking a module from the trailers or SPMTs and "lifting" it onto the deck, transporting, and then setting the module back on another set of SPMTs for the remaining transport to the site. Such vessels are an alternative to the RO-RO where the modules are not excessively large or heavy (less than 500 metric tons+/-), and the vessel can be brought alongside the MOF. In all cases, there are some important considerations that must be evaluated early in design to ensure a successful and seamless transition from the module fab yard to the transporting vessel and from the transporting vessel to another set of land transport trailers or SPMTs which will take the module to site.

9.7.3 Capacity of the Wharf or Quay

As mentioned in Chapter 5 under project drivers, the interface between the module fab yard and the vessel can be challenging and should be analyzed as part of the fab yard review. There is a certain depth requirement at the wharf edge or quayside that must accommodate the ballasting of the RO-RO vessel deep enough to align the vessel deck with the wharf height. There are also slightly different clearance requirements between the vessel propulsion screws or thrusters and both the bottom and the vertical edge of the wharf or quay. Since the actual module load-out is not a very fast process, accommodation of the timing and magnitude of the local tidal range is required. Even though the load-out is typically scheduled between a rising and falling tide, the vessel must have the ballast capacity to accommodate this rising or falling tide. Such details can be quickly worked out between the vessel owner and the fab yard but need to be evaluated early to ensure no last-minute additional ballast capacity or dredging is required.

9.8 Module Movement Considerations

At some point in time, every module will need to be moved—usually multiple times. It is important that the module be designed from the beginning for these moves. Typically, this is easily accomplished by providing a structural frame on the bottom of the equipment module or the lowest level of the pipe rack module. This will require that these bottom members be designed for the additional loads of the entire module weight and specifically designed for any local concentrations of the module weight in the areas of the actual support interface between module and transporter.

In addition to this design, early consideration is required in terms of exactly "how" the module will be secured to the transporters as well as how they will be secured to the vessel deck. Since the actual module support points on the deck must correspond to the internal framing of the vessel itself, this must be considered in the design and will be mentioned in more detail in Section 9.10, Shipping Considerations. But, even prior to this, the basic design of the module must provide access to both the lower structural members as well as potentially the outside of module's major structural members to permit either tie-downs, chains, welded brackets, or a combination of them to be installed after the module is completed. This requires that the design identify these areas such that piping, equipment, or E&I&C are not installed in or around them.

9.9 Module Responsibility Matrix

There is also the issue of exactly where each transfer of custody occurs. This is important since there are several different contractors involved in what may seem on the surface to be a relatively contiguous operation. This multiple custody transfer must be identified and agreed to early in the fab yard and vessel contracting negotiations. Consider the following example of the different custody transfers:

- **Equipment:** The project purchased the material, and the equipment is shipped to the fab yard. When received, the fab yard typically takes responsibility of the equipment.

- **Module movement:** The fab yard will continue with responsibility for the movements within the yard while building the module. But, upon completion, they may work with a third party logistics company, requiring this third party logistics company use the fab yard trailers and SPMTs but turning custody over to the third party logistics company at that time.

- **Module load-out:** The third party logistics company will organize and move the module from the quayside onto the vessel, at which point the custody transfers to the ship. The actual point of custody transfer varies but is defined ahead of time. During the short interval between quayside and vessel, all interested parties play an active role in ensuring their piece of equipment or property (quayside capacity/dock plates across gaps to vessel/SPMT maintenance and operation/placement on the vessel deck/tie-down and sea-fastening) is functional to make sure all goes smoothly.

- **Module tie-down:** Once on the vessel, the module custody transfers to the vessel owner, who has previously evaluated the proposed module placement on the deck in terms of vessel shipping stability as well as the specific hardware and method of tie-down to the deck. Typically this work goes back to the fab yard because the fab yard has the equipment (welding machines and materials) as well as the craft team (welders and fitters). A marine warranty surveyor and various insurance representatives for the owner/client, as well as potentially the contractor, get involved in the review and supervision to confirm the load-out and tie-down are implemented as previously agreed.

- **Transportation:** The vessel owner or captain has custody of the module for the ocean-going trip, releasing this custody at the destination where the custody transfers are reversed from the above load-out.

- **Offloading and trip to the site:** The third party logistics company (typically the same one that performed the load-out at the fab yard) supplies the appropriate trailers or SPMTs and supervises or performs tie-down and preparations for movement to the site.

- **Module set at the site:** The third party logistics company maintains custody all the way to the point where the module is finally set on its foundation, at which time custody is turned over to the site construction team.

9.10 Shipping Considerations

Early in the module analysis, the subject of shipping (how you get the module from the fab yard to the site) should be discussed. Many early design decisions depend on how the module will be moved between the fab yard and the site. For example, the acceleration forces on the module during shipment will be influenced by the type of vessel and location of the module in relation to the vessel's metacenter, location fore or aft of center, and the height of the module itself.

- For sea transport (see Figure 9.21), it will be a function of which portions of the world's oceans the module-carrying vessel will cross.

- For land transport (see Figure 9.22), it will be the road width and configuration, as well as the speed of transport.

9.10.1 Tying or Fastening the Module

As mentioned in the load-out considerations, the method of tying the module to the vessel must be flexible in terms of where the module is tied to the deck since the deck is strongest at the points where the deck is supported by the vessel's internal bracing.

Figure 9.21 Sea transport, ready to ship.

Figure 9.22 Land transport, Wyoming, USA.

Two terms are used when talking about how the module is "fastened" to the vessel deck: grillage and sea-fastening.

9.10.1.1 Grillage

This is the collective set of mats, timbers, stools, or partial beams on which the module actually rests. The purpose of the grillage is to maintain the space between the bottom module steel and the deck so that the trailers or SPMTs can be moved out after positioning the module on the vessel.

9.10.1.2 Sea-Fastening

This is the actual wire rope, chains, welded steel brackets, etc., that physically tie the module to the vessel. In the case of smaller modules, lashing with wire rope or chains may be adequate. For the very large modules, the acceleration s forces must be resisted by more substantial welded brackets and braces.

The design can be simple (Figure 9.23) or complex (Figures 9.24 (a) and 9.24 (b)), but in all cases must adequately restrain the module on the deck and tie the parts of the module that will adequately transmit the module accelerations to the main vessel structural trusses.

So, very early in the design, sections of the module lower steel must be kept clear of any equipment, piping, electrical, and instrumentation. In some cases, because of the need to place the sea-fastening and grillage on the deck, specifically at the strong internal trusses of the vessel, it is

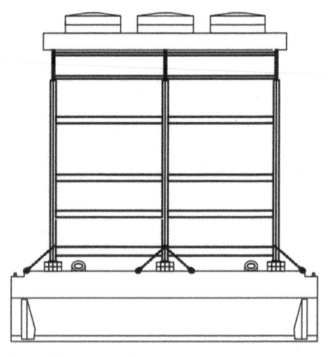

Figure 9.23 Sketch of simple timber grillage with chain sea-fastening.

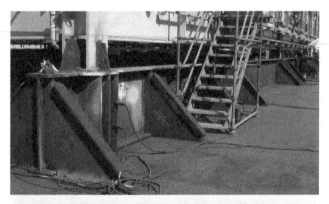

Figure 9.24 (Top) Complex structural beam grillage; and (Bottom) tubular sea-fastening combinations.

best to leave the entire lower module beam and the lower area around each column free of appurtenances, piping, and other things that might get in the way of the tie-down.

The more complex the grillage and sea-fastening, the more costly it is to fabricate and install. But, in addition to making both seaworthy, it is important that both be designed so they can be quickly installed and tied or welded down after rolling the module onto the deck as well as quickly removed once the vessel arrives at the wharf or quayside destination. A poorly designed combination of grillage and sea-fastening can take longer to both install and later remove, lengthening the total time to load and unload. This is a potential concern where the loading or unloading has to deal with high tides. Also, if the vessel is delayed beyond the contractual duration(s) for loading and unloading, there are additional demurrage charges that can be expensive.

9.10.2 Movement to Site and Hook-Up

9.10.2.1 Load-Out

Moving the module around the assembly yard and loading it onto a truck or vessel for shipment are typically part of the fabricator's scope of work and, as such, will be considered by them in their scope of work. But, it is important that this be discussed with the proposed fabricator, as it is one of the critical activities that the module fabricator must perform. They will have a specific scope and point of custody transfer that need to be identified and agreed upon. This is one of the standard negotiations with a fabricator and, as such, is not typically missed.

9.10.2.2 Offloading

However, the entire trip from the point where the module is offloaded from the vessel up to the project site is a part of the project to develop and can be problematic if not adequately planned. The module route will require much more coordination, from the destination MOF (which in many cases is a public / commercial wharf, with its own set of coordination requirements), along the public roads (with all of their restrictions from road width, utility, and power poles and lines, all the way down to fences, ditches, retaining walls, abutments, etc.), up to the site itself (with the appropriate road modifications or additions) and finally within the site (via a temporary, but purpose-built HH access path that is designed to provide a clear path to the multiple module setting area).

9.10.2.3 The Heavy Haul (HH) Route Survey

It is this part of the module effort that should be developed in the initial module transport analysis previously mentioned as one of the very first activities in the project planning of a module job—that being the determination of the maximum shipping envelope (from the receiving MOF to the proposed project site). The heavy haul route survey or study can be as simple as identification of a route and a few measurements indicating where potential pinch points are. But typically, as indicated above, the path from MOF to the site is complex enough that this survey or study becomes a formal multi-page analysis complete with routes, alternatives, and annotated photos of each potential move restriction (along with suggestions on how best to resolve them).

Critical to the success of the project is finding a knowledgeable company to perform the HH route survey. Typically, a local company familiar with the area, roads, regulations, permitting, local authorities, etc., will provide the most complete route analysis and best option(s). However, for unique situations, very tall towers or heavy equipment, and very large modules, the contracting of a more international logistics company may be the better choice due to the company's familiarity and experience with the more unique transport configurations and equipment that might be required for these special long, wide, high or heavy loads. In most cases, these international logistics companies will have one or more local offices also employing local personnel who will be able to provide the specific transportation details regarding regulations, permitting, and required approvals necessary to ensure the module movements will not be held up due to "paperwork."

9.10.2.4 Logistics Coordination

It is advantageous to have the same logistics company coordinating BOTH the load-out from the fab yard as well as the offloading, movement to the site, and final module setting. Yes, this seems like it should be an obvious decision. But, when working with multiple module fabricators, numerous heavy equipment suppliers, as well as separate structural steel and pipe spool fabrication contractors, all of which have their own ideas regarding how they can utilize their own equipment for movement and shipping, single-point coordination of all these moving parts becomes important.

For example, with slightly different trailer heights and capabilities, it is important that the gap provided between the ground/deck and the lower structural steel load-carrying beam will accommodate all of the trailer and SPMT types that are planned to be used in execution.

Most large EPC firms will have their own internal logistics group that is responsible for this overall project coordination. They will typically work with the vessel owners, heavy haul (HH) and heavy lift (HL) companies, setting the groundwork and rules on scope splits and responsibilities to ensure that there are no gaps in the movement sequence from fab yard to final setting.

Also, many of the larger fabrication yards, as well as the shipping companies, provide integrated logistics services. For example, the fab yard may also have HH and HL capabilities as well as ocean-going transport capabilities. This provides additional revenue streams for these companies. For the project, it also provides an option of limiting the number of different sub-contractors involved in the module movement. Limiting the number of custody transfers limits the chances of having overlap or gaps in the module movement as well as limits the potential opportunity for miscommunications between contractors and the resulting opportunities for scope change, change orders, and their resulting additional costs. Anything that reduces the number of contractors and their coordination should be considered favorably in the early project coordination.

A fabrication yard that also has responsibility for shipping will make sure that the vessel will be available when the module is ready to ship. Delays in module fabrication (assuming these are not caused by the project) will be the responsibility of the fabrication yard to correct. In general, combining these two scopes is desirable, assuming, of course, the project is OK with any perceived additional risk of awarding both fabrication and logistics to the same company.

9.10.2.5 On the Site

The final coordination will take place within the project site. While the HH contractor will work to move the modules and other equipment as scheduled to the site, they will not take responsibility for any problems with getting the modules to their respective setting locations due to construction site activities. They will be relying on the on-site construction management team to coordinate all the activity within the site (excavations, pile driving, and any early foundation and equipment setting) to ensure that the path to the respective module setting locations is clear.

This can become a challenge, especially when module equipment is late, the module itself is late, or there is a change in the path of construction or sequencing at the site. During the initial plot plan layout, the plot plan path of construction should be developed, providing as much flexibility in module setting as possible without creating excessive cost implications. For example, a site with the ability to install main pipe rack modules from either end offers opportunities to make adjustments in the rack setting sequence that may benefit the adjacent site construction hook-up.

9.10.2.6 Single Weld Hook-Up (SWHU)

Historically, when building a plant, due to small differences in the setting of major equipment, connecting pipe spools were designed with 3 degrees of freedom. When the equipment was set and the piping to be attached to it was in place, final measurements were taken between the equipment and the proposed connection, and these "tie-in" spools were fabricated for an exact fit. Besides requiring exact measurements, it also required two connections be made—one at each end of the tie-in spool. For many of the industrial processes, this amounted to two welds/pipe.

With the advent of the 3D model, laser scanning, and distance measurements, the construction industry is technically capable of building a module to "exact" dimensions (Choi et al., 2020), even taking into account the impact to these dimensions from the thermal expansion of the piping due to temperature differences between the fabrication yard and the project site. (We will explain in more detail in Chapter 12.) With this accuracy, the connection between two modules/a module and a piece of pipe-rack pipe/a module and a pipe connection to a tower can be made with one weld—resulting in a single weld hook-up (SWHU). This capability provides the opportunity of designing these modules for SWHU (Figure 9.25).

So, what exactly is a SWHU? It is more than just the one weld that makes the connection between two modules

Figure 9.25 SWHU example alignment.

or a module and a piece of equipment or piping. It is an entire philosophy of design, fabrication, testing, pre-commissioning, and quality control to be able to use a single weld to connect two previously tested and qualified areas of the project plant.

Why the interest in the SWHU? As mentioned, the typical connection between the module piping and site piping involves two connections (weld or flange) and a short piece of pipe. With some advance planning and use of the SWHU philosophy, this can be cut down to a single weld or flange. That is a 50% saving in labor and consumables. While that may not seem like much, putting it into perspective, on a large project we were associated with, the total number of SWHU pipe welds (sometimes referred to as closure welds) was over 8,100! That meant that this was savings to the project of 8,100 extra welds! Because of the requirements in terms of design, planning, and procedure approvals, it is important that the decision to use the SWHU process is taken early in the design process.

9.10.2.7 SWHU and Owner and EPC Coordination

Of special importance is the approval of the proposed SWHU procedure by the owner and construction teams. This is important because the whole reason for using the SWHU is to be able to test the entire piping system on either side of the closure weld with the project standard NDE or hydrotest at the module fab yard. But, along with this testing, there must be all the paperwork (appropriately signed) required to document that both sides of the proposed closure weld piping have been tested and verified. Included in this documentation is confirmation that both pieces of piping are ready for the closure weld as well as the details for the actual closure weld and subsequent NDE.

We really are making a big deal trying to explain the importance of developing procedures that have been approved by project management at a level that causes no second-guessing. There have been too many projects where a half-baked SWHU procedure makes it all the way to the site, only to have a quality control or quality assurance representative disagree with the procedure and require re-testing, re-hydro, and all other sorts of re-work, not only negating the potential savings but creating additional costs and schedule impacts that were never there. In the end, the only way to be successful is to document—document—document and get written approvals from a level of the project that is irrefutable.

9.11 Construction Considerations

The construction team needs to understand the module schedule and modules.

9.11.1 Understand the Schedule

This is different than a stick-built schedule. Construction does not have the typical latitude in approaching the site work any way they want to. There are specific activities that must be completed in a specific sequence to permit the modules to be moved into place.

9.11.2 Understand the Modules

The construction team must work with the module fab yard management to understand what will be sent to the site, how it will be sent, and what parts may still be unfinished.

For example, in the past, one construction manager commented when one of the first modules arrived aboard the ocean-going vessel: "I had no idea they were that big!" Such surprise could have been avoided by agreeing to make the visit to the fab yard instead of declining it on several occasions. This comment was followed weeks later with a second acknowledgment upon seeing how easily the modules were set and how well they hooked up to the equipment and piping at the site: "I really wish more of this plant had been modularized!"

Sadly, unless the construction team has worked with modular construction before, they will not fully understand the benefits these modules provide to their site. They will continue to try to run the site construction effort the way they ran a past stick-built job.

How to resolve? If you cannot get a module-savvy construction team, do the next best thing—make sure that they are involved in all the upfront design, scheduling, and procurement developments. Only then do they have an opportunity to understand the benefits of prefabrication and the tight scheduling constraints that must be followed to make the project a success.

9.12 Summary

In this chapter, we explained the different module considerations by the project group. The subjects discussed were: Engineering, Schedule, Procurement, Sub-Contract, Fabrication, Completion/Testing/Prep, Load-out, Module Movement, Shipping, and Construction. These points will help companies to successfully implement modularization if they follow our recommendations.

References

Choi, J.O. *et al*. (2020) Innovative Technologies and Management Approaches for Facility Design Standardization and Modularization of Capital Projects. *Journal of Management in Engineering*, 36(5), 04020042. doi: 10.1061/(ASCE)ME.1943-5479.0000805.

Han, B. and Leite, F. (2021) Measuring the Impact of Immersive Virtual Reality on Construction Design Review Applications: Head-Mounted Display versus Desktop Monitor. *Journal of Construction Engineering and Management*, 147(6), 04021042. doi: 10.1061/(ASCE)CO.1943-7862.0002056.

chapter 10 A Practical Module Development Process

This chapter is where we, the authors, get to expound a bit on our learning experiences, even if many of these are the results of painful missteps, wrong turns taken down blind alleys, and too many chases down (dead end) rabbit holes. But this chapter is also interspersed with the information provided (sometimes almost as if by divine guidance) from a smarter and more experienced coworker, and one willing to take the time to extend a helping hand as we struggled while simultaneously implementing and learning. AND, by no means do we claim to know it all. In fact, the almost scary reality is that every new module job tackled comes with new learning experiences, nuances to project execution that get added to the collective knowledge we have (and try to remember), hopefully providing for the development of more complete and accurate responses to later related issues when asked (assuming, that is, we ever even get asked!).

Modularization is not "rocket science"—at least not in our opinion. As with most great concepts, it is based on some simple tenets, including planning, organization, teamwork, singleness of purpose, and commitment, just to name a few. Modularization is also like the over-used Harley Davidson phrase—"it's not the destination, but the journey . . ." It is not just knowing the facts and figures on how to do it correctly; it is the actual process of fusing a group of like-minded individuals into a team and then working together to figure out a way to get the project done. Only as a result of this collective effort or "journey" do we actually develop all the "neat little things" that help us make the journey or project easier and the final destination a true accomplishment.

So, a bit of advice before we get too far into this. Share the little nuances you know, learn, or may have been told. They are of little use by themselves (and don't do any good stuck in your head, never to see the light of day). It is only in the context of the collective project that they provide benefits and can be built into and on top of the current experiences and knowledge to expand and progress the greater module good.

So much for "waxing eloquent" on the philosophy of sharing . . . Both of us have found the more we share what we thought was perhaps somewhat "unique or special," the more we found out it was closer to common industry knowledge, with others already sharing how they had enhanced or modified it to fit their specific needs. Then, based on your knowledge of the subject, we will let you decide whether this is a "typical" or an "atypical" example.

10.1 Introduction

This chapter goes into more detail on some of the potential questions, early resistance, and opportunities to communicate the benefits of using modularization as an alternative to stick-built (before the opportunity is "submarined" by well-meaning management). Its target audience is the "maybe not so experienced" project engineer/manager who may end up being assigned as "module coordinator." Its goal is to try to identify and provide input in terms of responses, grounded in our experiences, to some of the questions that come up during early study work as well as once into the actual execution of a module job.

Much of our input is aimed at the reluctant management audience and module detractors, as it takes only a handful to create an environment where module project success will be difficult, if not impossible. And while each project and set of managers are different, many of the issues are similar. So, it is hoped that this section will help highlight some of the more common stumbling blocks encountered on your way to the successful implementation of a module program.

Oh, yea, be forewarned—while we followed a logical sequence in the development of this book to explain the evolution of a module job up to this point, the writing in this chapter seems to have followed a bit more stream-of-consciousness style than anticipated, as one thought and response feeds other related ones, all of which are important at different times. So, if you can accept the somewhat wandering thought process and keep an open mind, this chapter will provide answers to what many of you need when trying to get a module project started and moving.

10.2 Initial Project Analysis

How does all this start? Someone suggests a project be modular. Typically, this starts with a comment that the owner/client makes, or they simply ask, "Can you modularize my project?" The obvious answer from your company business development representatives' standpoint is "Yes!" But to make any project a modular success, a quick reality check should be made (just prior to responding with such enthusiasm).

10.2.1 Reality Check Questions

It is suggested that the following self-examination of both the owner/client motives and the project potential be initiated.

- Does the owner/client expect cost and schedule savings with modularization?

- Does the owner/client have an idea about how to modularize the project (or are they simply asking a question)?

- How far along is the owner/client's design? (Hint: Anything more advanced than early pre-FEED should start raising red flags.)

- What is the owner/client's experience with past module jobs? (Are any of them positive?)

- What is the owner/client's attitude toward a different approach to the project planning efforts?

- How willing is the owner/client to provide additional funding prior to FID?

- How familiar is the owner/client's team with participating in a modular project?

Notice something that all these reality check questions have in common? If not, take a closer look: they all involve the owner/client input. As we have mentioned in Chapter 7 on the critical success factor, the owner/client is responsible for the majority of the critical decisions on a module job. Reluctance, indecision, or resistance to change are not a good initial sign—one that must be caught early, explored, and resolved through education.

Of course, it is assumed that you are not facing the same lack of (module) understanding within your EPC organization. But, if you are, the same approach (and questions) should be used within your organization! Of course, we suggest you approach your organization with a bit of tact since a poor approach and delivery to an overly sensitive upper manager could be a career-limiting move for you as the younger member on the proposed project team.

If there are weaknesses in the owner/client knowledge base (or even in your own organization), it will be your first duty to figure out where these are and, based on that, how to address them. For this lack of understanding or lack of commitment, a little bit more investigation is suggested to determine what part of "modular" they do not understand (along the lines of the similar familiar warning—what part of "no" do you not understand?—but hopefully not so blunt!)

10.3 Early Discussions

Suggest the following thoughts/ideas (call them module tenets) be woven into subsequent conversations as soon as you identify the particular reason(s) for hesitation or a lack of understanding. The goal of these introductory and early follow-up discussions is to identify any common misconceptions held regarding modularization and replace them with facts and updated information (so the team can start with a more or less neutral opinion of module opportunities and benefits versus potential downsides).

These early discussions are important because, like the module "Perfect Storm" from Chapter 7, almost everyone has a module horror story they will share with you first before sharing any good project module experiences (assuming they had ever even had some).

10.4 Module Tenets

Listed below are module tenets (commonly held views and beliefs on modularization). These are by no means all of them, but they are the ones that quickly popped into our minds as we asked ourselves the question:

What is it about modularization that EVERYONE should know?

10.4.1 Definitions

- **Modularization and pre-assembly are essentially the same thing.** Pre-assembly is just a simpler form of how to move craft work-hours off site. This combined concept definition is crucial as it opens up many opportunities for alternatives to the traditional stick-built execution plan.

- **A module job consists of a mixture of big and little modules.** Even the biggest module job will probably have several "small" (less than 100 ton) modules and numerous pre-assemblies.

- **There is no "minimum distance" for moving off site.** Moving work-hours off site can be as simple as pre-assembly on land directly adjacent to the project site. Anything that creates a way to build something "cheaper" is fair game. As an example, this can be building something at grade next door if this produces a finished product with the increased productivity of working at grade and by doing so allows more space for craft in the on-site more critical areas.

10.4.2 Boundaries and Limitations

- **Not everything on a project can be modularized, but every job can benefit from some sort of pre-assembly.**

- **Modules improve access.** A properly designed module will IMPROVE access to equipment for O&M activities, not limit it. Think about it—most modules provide a third dimension (the vertical) in which to support equipment/piping/E&I&C as well as maintenance appurtenances. By expanding the width to accommodate access ways, the combination provides improved overhead support opportunities as well as easier access horizontally.

- **Modules reduce safety hazards.** A properly designed module should minimize or eliminate safety hazards, not create additional ones. Again, the overhead features of a volumetric module provide opportunities for proper supports, proper access, and proper protection.

- **Modules are not "cramped."** A properly designed module is not "cramped" in terms of spacing. There is no reason to make the land module overly constrictive. (Do not confuse onshore and offshore module restrictions—offshore modules are limited in size by the supporting footprint of the jacket, semi, etc. The land-based module has no such restriction.)

10.4.3 Cost

- **Early module planning does not cost more.** Early module-based planning (the feasibility and selection phases) will cost the project no more than conventional stick-built planning. The actual work performed in a project's assessment and selection phases is essentially the same whether you have a modular or stick-built execution plan in mind. The difference is that with a modular execution plan, you will need to start planning a lot of actions earlier. But early planning makes for a better end product, so what is the loss? Even if the project decides against the module philosophy, the earlier planning efforts required will have provided an early kick start to that project and will typically leave the project better prepared for upcoming activities in later phases.

The complete modular job will increase the typical 6–8% engineering job portion of the total installed cost (TIC) project cost by approximately 5–10%. This varies from project to project but is typically due to the extra support costs primarily for the additional work associated with the module fabrication, shipping, and setting efforts and the extra coordination of these activities. But this is often offset by the benefits gained from the earlier planning in terms of design and procurement—more time to do it right (the first time) than do it over (at the last minute).

10.4.4 Impacts

- **Execution Plan Differences.** The module job impacts "all" aspects of project planning and execution. Many are very subtle, but timing is critical. For example, everything on the module will need to be "completely engineered" before sending it to the module fabricator (to avoid expensive changes at the fabricator's yard or carry over to the site). This means you cannot allow your favorite construction manager to "re-engineer" the on-site execution of the facility—there is much less flexibility with respect to timing.

- **Critical success factors (CSFs).** These are actions that are critical to the success of the module job. They should be identified and embraced by the team. Hint: These CSFs are all really common-sense actions and attitudes that will make any job (stick-built or modular) more successful. The reason why these are so important with a modular execution plan is that there is much less latitude for variation during the construction sequence (the path of construction): timing is everything.

- Both of the above concepts and their impacts on the project design and execution need to be common knowledge for all members of the project team (and embraced by them).

10.4.5 Experience

- **Every company needs a module "expert."** The contractor and owner/client BOTH should have a module "expert" on the team – identified and available for early consultations. If no one has been identified, make sure one is found for each and make sure each organization brings one on. The time to bring these on is as early as you can because even in the opportunity framing phase (FEL-0), there are decisions to be made that impact potential module outcomes.

- **Every company needs to be module "savvy."** Contractor and owner teams BOTH need to be module "savvy." Inadequate technical background on either or both sides creates project delays while working to "educate" one or the other on why specific activities are warranted either earlier or in a different order than what they are used to.

10.4.6 Commitment

- **Total commitment is required.** Contractor and owner teams BOTH need to be completely committed to the module solution. A single detractor of the proposed module solution will seriously endanger the chances of success. The team's energy needs to be spent on developing solutions, not debating who is correct or explaining and defending the approaches taken (which will, by definition, be slightly different).

- **Flexibility is required.** The owner/client needs to be willing to work with the slightly different priorities and sequencing of early work. This always goes smoother if BOTH teams have a module-savvy internal resource they can confide in and work with.

- **Collaboration is required.** Finally, is the owner/client (as well as your own project management) working internally as well as externally with each other in a collaborative manner rather than taking a transactional approach? Remember, many activities that typically are performed a certain way will need to be adjusted a bit with a modular execution approach. The entire team will need to be flexible, strive to understand why these (execution plan) differences are necessary, and then accept and support them.

10.5 Project Drivers

After the preliminary alignment session(s) on module philosophy and a discussion (or two) on the above pertinent module tenets with the goal in mind of possibly leveling the module playing field discussions from "everyone's worst nightmare" to "a concept that might have some potential," the real reasons why the owner/client wants a modularized project should be ferreted out. After addressing (and dismissing) some of the more "universally shallow" reasons (e.g., "Modularization is something everyone is doing these days" or "It is something that is guaranteed to save us money" or even the more philanthropic view that "It is the right thing to do"), there needs to be a discussion on why this particular project should be modularized.

10.5.1 Cost versus Schedule

While there may be some "lip service" to better quality, safety, and more local content, the two big elephants in the room are always *cost* and *schedule*. The owner/client will be quick to say that both are equally important. And while both are important, one will be more important than the other for economic or scheduling reasons that the owner/client may not wish to share with the EPC contractor. In any case, it is important that the owner/client be coached into at least identifying (in their own mind) which one is more critical to them in terms of net present value (NPV), rate of return (ROR), total installed cost (TIC), or whatever metric the owner/client equates project success with. Once the owner/client has done so, it is important that the owner/client communicate which of the two is more critical to the EPC contractors. The owner/client may not want to disclose or explain why one is more important than the other, but the owner/client needs to clearly communicate which one is their priority, as there will be a slightly different approach to the module project planning depending on which is deemed more critical.

There are several ways to approach this conversation. We typically start with a plan to identify some significantly unacceptable cost increase and compare it to an equally significant schedule increase and offer to help the owner/client walk through the exercise. At some point in the conversation, as you continue to adjust the two increases higher, one will become unacceptable (for internal budgeting/NPV/ROR or other reasons, maybe not shared). The owner/client will indicate which one cannot be tolerated. After a quick confirmation, you move forward with the choice provided.

You may get push back from the owner/client on why such a decision is required. We suggest the following line of reasoning for your potential response: A modular execution philosophy is developed as a team effort. The modules are typically on the project critical path. There are many different alternatives in terms of module configuration—mixing and matching the equipment within

the modules as well as varying the number of modules in the project. Some of these modules will be impacted by the equipment in them. So, it is important that they be examined based on the potential impact to either "cost" or "schedule" in terms of late or missing delivery to the fabrication yard. Knowing whether cost or schedule is more critical can influence whether you include that piece of equipment or kit on the module (in order to maximize work-hour removal from site and dollar savings) or leave it for separate shipping and setting at the site (increasing the extra work-hours associated with a project site hook-up, but saving schedule by getting the now smaller module to site earlier).

After all of this, it is still very likely that the EPC contractor will not be told why which one will take precedence, but, in most cases, they can figure it out. However, if the true project economics are not shared with the EPC project contractor, you now have the owner/client's marching orders, and you can effectively plan where there is a potential trade-off between cost versus schedule.

10.5.2 Other Drivers

After the cost vs. schedule decision—go ahead and breathe a sigh of relief but keep your celebrations short! You now need to continue to work on the identification of the rest of the project drivers from the owner/client's perspective.

Many of these have been mentioned in Chapter 5, The Business Case for Modularization. Getting as complete a listing from the owner/client as possible becomes essential when evaluating module options and total percentage to modularize. Depending on importance, these other drivers may push the project to either more modularization or less.

For example, limiting impact on local community or impact on nearby land would drive a modular solution that results in the movement of the most construction work-hours possible off site to reduce the workforce on site. Conversely, a local community that wants local craft employed will push for work they can do locally and may shift the total number or size of modules to accommodate the specific skills of the local craft. Both are correct answers but one is a better option for your particular project and should be the goal.

Safety/quality are always on the list and are a given when it comes to project drivers. But what parts of past projects have the owner/client had issues with on either safety or quality? Poor welding quality/high reject rates on alloy materials may suggest that the project should maximize specific process units so these units could be built in an enclosed area of a fab yard where such alloy welding is their specialty. The same concept could be applied to E&I&C if local craft quality were an issue.

Are there parts of a process that need to be designed and assembled in specific countries (or conversely, are there some countries that the owner/client wishes to avoid)? Do customs clearance and import duties have an impact on where the units get built?

All project drivers can impact the modularization scope and should be identified upfront to ensure the best fit in terms of the final modularization conclusion.

10.5.3 Owner-Furnished Equipment

Sometimes an owner/client can even be too helpful. This may seem counter-intuitive, as help from an owner/client in terms of taking the risk of purchasing critical equipment seems like one less thing for the EPC contractor to worry about. But, what if the owner/client contracted the purchase of a major piece of equipment that requires extensive assembly at the site? What about the equipment that has been purchased at a significantly reduced price (significant project savings) but with a delivery that will not meet the needs of the module fabricator for inclusion in a specific module? Finally, what about equipment purchased with no accommodations in the contract for equipment support because no thought was given to the necessary long-term maintenance for the duration of the equipment just sitting in the module or even a third party representative at the module fab yard for assembly, preservation, and maintenance?

All such good intentions can be extremely challenging if the results do not fit the project module schedule or scope. Of course, in some cases, the damage has already been done by the time the owner/client brings you on board as the EPC contractor. In other cases, it may be your procurement group that continues to reproduce the same purchasing contract details that do not

explore the options of pre-assembly or purchase of critical engineering earlier, or even options on early delivery or split delivery sites. In either case, it becomes very important that these small details get addressed upfront and BEFORE they become bigger stumbling blocks when trying to develop the module equipment compositions. This is an excellent segue into the next subject of discussion: the module team.

10.5.4 The Module Team

10.5.4.1 The Module Coordinator and the Business Development Team

The module team starts out as a single person: a module "representative" or module coordinator. Very humble beginnings, so when should this single solitary module person be brought on board? At the very beginning of a prospect pursuit. The module coordinator should be at least a part-time representative on the business development (BD) prospect pursuit team.

But this seldom happens. Why? Typically, either the BD or pursuit team management doesn't understand or appreciate the opportunities that can be identified via an early conversation with the owner/client on alternative project execution philosophies (e.g., modularization and pre-assembly). Or, even more embarrassing and irritating, they assume it won't come up or, if it does come up, they assume they have the background knowledge or ability to cover any such questions.

Many times, because of budget constraints, the BD team assumes this "extra" part-time module support is not necessary. Unfortunately, there are very few people, if any, in BD who really understand the ramifications of the decision to modularize. The only redeeming grace for the uninformed BD person is that they are "lucky" enough to be working with an equally uninformed owner/client.

Attempts by BD to "bluff" their way through a detailed discussion with an owner/client who happens to be knowledgeable in the subject will only lead to misunderstandings at best and a disgruntled owner/client and a lost opportunity at the worst. It is best to make sure the prospective project is addressed with the appropriately correct module understandings.

Why are we so emphatic on getting module representation this early? First impressions are critical. Owners/clients have become very sophisticated in terms of detailed knowledge on many fronts and can easily spot a weakness in technical ability. Companies have figured this out and usually bring their process experts to answer the technical questions, and their expert piping leads to respond to equipment layout questions. But what about the questions on construction, path of construction, modularization, or standardization (and now even Advance Work Packaging or Lean concepts)? With most of the project costs part of the field execution, these topics should be addressed by someone who has actually been out on a project site more than just to see it. The goal of the BD pursuit team is to impress the client with their collective abilities in terms of the total job, including the module and other field expertise, as quickly as possible.

In addition, you do not want to spend the next several meetings with the prospective owner/client trying to politely explain why what the BD person said was "not exactly correct." You know this is coming when you hear comments from BD such as: "Wish we had you here to answer this or that question" or "we didn't realize that question was so involved."

Why so much emphasis on the BD role and first impressions? (Sorry, lots of skeletons in that closet.) Too many BD teams run on such a tight budget; they only have enough budget for a very few to be able to face the client, with everyone else "on-call."

10.5.4.2 Prospect to Project: Module Team Growth

Somehow, the prospect pursuit was successful, and the company now has a brand-new project. The BD pursuit team has completed congratulating themselves and has now unceremoniously "handed" this project (along with all their overinflated promises to the client) to the project management team, who now must execute it. But, hopefully, by this time, project management understands the need for a module presence and expertise and has identified a module coordinator, or on a very large project, an entire module team, and set one up in the organization. The module lead/team and their importance were previously discussed in Chapter 6 (which the reader might want to review before going further).

But, just like the BD team efforts, it does the project a little good if the module team lead is not included in all the early project planning and discussions with the owner/client. Just as seemingly simple efforts by the owner/client to help in purchasing long-lead equipment can adversely impact the optimal module solution, so can an uninformed but well-meaning EPC management team make decisions that will cause additional problems in achieving the optimal module solution.

As a first step, the module team lead must quickly identify their counterpart on the owner/client side and begin early discussions with them to understand any strengths to be counted on or weaknesses to be understood and dealt with.

Such early engagement will go a long way in developing the mature respect and rapport that will be valuable later in the project when trying to get variations in the project planning (as a result of the module approach) approved. In addition, understanding any weaknesses or gaps in modular knowledge creates opportunities to "share" with your owner/client counterpart and provide any needed early heads up on what might be coming in terms of changes to the owner/client's standard project protocols.

At the same time the module team lead is getting to know their owner/client counterpart, they must begin recruiting "willing" individual representatives from each of the major groups within their side of the project. As previously mentioned, these will be the eyes and ears of the module team as they each work within their respective groups.

Why is it important that there are representatives from each of the major project groups in the module team? Remember the 107 execution plan differences that were summarized under 21 separate topics in Chapter 6? These execution plan differences impact *ALL* of the project groups. Since the module team lead cannot be "everywhere—all the time," the team leader must rely on members within each of these major project groups to keep an "ear" out for any project planning that might impact the module efforts, either positively or negatively.

Of course, with such an agenda for these module team members, there is usually one or more group managers who may get offended with your proposed attempt to get "one of their own" to, in effect, "spy" on their group.

Think about it, you as a module team leader are requesting someone to work part-time for you, typically without chargeability, with the goal of "ratting out" the actions of their discipline or group leader for doing something not in line with the module agenda.

This is where the module team member's great skills at "schmoozing" come in. Do you remember that course in college? It was the one that most engineers avoided like the plague because it involved personal interaction skill development, something most of us were not born with and had trouble even getting below average at. But seriously, this is an important part of being a module team leader and should not be taken lightly. As mentioned, it is critical that the module job has everyone marching in the same direction and being enthusiastic about it. As the module team leader, in some cases, you may be the only one on the project that understands how the groups must adjust actions and deliverables to make the project a success.

So, with a unified module team, supported by each member's discipline lead/group leader, we can now start talking about what needs to be done on the project itself, starting with the modules themselves.

10.5.5 How Many and How Big?

We started this chapter by warning you that the path from the first effort to the completed project may not be a straight arrow shot. In fact, for some projects, the explanations and path we have provided thus far may have put the "cart before the horse." By now, someone on the project management team is probably asking: "Has anyone determined the number and size of the modules yet?"

Even if the project has been provided with a module listing with some amount of detail, it is important that a reality check be made on whether this listing of modules is indeed the optimal number and size.

Every EPC company that has performed module work in the past has developed a way to identify the best module case scenario for a given project. However, at some distant time in the past, each of those companies had to start from scratch. So, for discussion purposes, we assume that the contractor has not made such an analysis, and their company is not familiar with the necessary

steps to develop such an analysis. So, what we provide is a *five-step module method* for such an analysis.

10.6 The Five-Step Module Development Process

The typical module development process includes numerous separate but related efforts that must be simultaneously developed or considered in order to produce a suggested module program that will be best suited for a specific project. Key among them are the following (in the authors' humble opinion):

- A way to compare the various module size and shape options, from very small to very large, to find the size that is "just right."

- A way to select the best fabrication facility from the wide range of fab yard sizes and types available worldwide, and further, to determine how each will impact the overall project cost.

- A way to initially define the various sizes and shapes of the individual modules within that optimal size range.

- A way to subsequently further optimize the module configuration.

- A way to incorporate them into the schedule.

- A way to ultimately compare their cost versus the traditional cost of stick-built construction.

With a bit of rearranging and consolidating, one can see this effort follows a basic five-step developmental pattern:

1. Determine optimal module size, specific to a project.

2. Develop specific modules around this base size.

3. Optimize individual modules based on further detailed analysis, path of construction, and other owner desires and preferences.

4. Determine best schedule fit.

5. Develop cost comparison of this best fit set of modules.

So, how does such an analysis work when your company is working with module job Serial #1 (along with creating the first module they have ever looked at developing in lieu of stick-building)?

Review the module project drivers previously mentioned in Chapter 5, The Business Case for Modularization, including the 13-step decision process. This is the theory behind the practice and includes a brief description of the tools that can be found and used in this effort. Remember that in almost all cases, the main driver for modularization is the cost difference between what it takes to complete 1 unit of work at the project job site versus what it takes to complete that same unit of work at a module fabrication yard. This economic productivity ratio (or EPR), a phrase coined by CII in RT-283, has been included in a discussion in Chapter 5. It is the identification of this difference in cost between doing a unit of work on the project site versus performing it in an offsite fabrication yard that provides the necessary additional funds required to support all the little "extras" we see in a modular job, such as extra engineering engagement, extra structural materials, extra fab yard supervision, extra shipping costs, extra heavy haul transport, and extra setting and hook-up costs. And, if everything goes as planned, this EPR provides the extra cost savings by going with the modular solution, all of which comes along with schedule savings.

10.6.1 Step #1 The Stick-Built versus Module Comparison

10.6.1.1 Starting with Nothing?

The first goal of the 5-step module analysis is to determine just how much difference there might be between the conventional stick-built and the modular approach. As mentioned in Chapter 5, this can be approached in terms of three options:

- small (or low) percentage modularization
- medium (or average) percentage modularization
- large (or high) percentage modularization.

Good rules of thumb for what these percentages would look like are: *10–30% versus 30–50% versus 50–80% of on-site total direct field work hours being moved to the offsite fabrication yard.*

To come up with an idea of the potential cost savings, the following simple comparison can be made. Typically, some idea of the total construction work-hours required

for a specific project is available or has been developed. Along with this information is some determination of the cost of labor as well as the craft productivity factor for the project site. That covers the stick-built costing needs. The bits of information missing are the identical factors for the module fab yards of interest: the craft all-in wage rate and productivity factor.

Providing examples of both of these will do nothing but get the authors in trouble and date this publication. In addition, there are various publications (some available for a small fee) that will provide all these statistics by area, region, or whatever split is desired. But we think that there is some good benefit in at least providing some very general boundaries for each of the following: (1) cost of labor, (2) all-in wage rate, and (3) productivity so that the casual (or first-time) reader can at least get grounded in some currently realistic ranges in order to put all the other differences into context.

First, some definitions.

10.6.1.2 Craft Wage Rates

What the skilled worker actually gets paid. This can range from $10/hour or less in developing countries to $25–35/hour in the US to even much higher in some places in Europe and Australia.

10.6.1.3 All-in Wage Rate (AIWR)

The sum of all the costs associated with paying the skilled worker the money they take home. It includes, not only the pay they see on their paycheck, but also all the other costs associated with being supported on site—things like: taxes/mob & demob costs/local transportation/small tools/camp costs/contractor's field management/home office management/PPE/consumables/construction equipment/materials handling equipment/QA & QC/and contractor profit.

Sometimes, in lieu of trying to detail all components of the "all-in wage rate," one can estimate it by using a multiplying factor on the actual craft pay rate to get the total cost/work-hour. Such a factor could run anywhere from 3 to 5 times the salary the craft worker is actually getting paid (i.e., what is actually seen on the check stub). So, the actual costs/work-hour of work (AIWR) can grow to totals of what some typically see when talking about the different areas of the world:

- USGC: $90–120/hour

- Europe: $80–100/hour

- Middle East/Africa: $30–40/hour

- SE Asia: $20–30/hour

- Far East: $40–60/hour

- Australia: $120–200/hour

10.6.1.4 Productivity Factor (PF)

This is a number used to compare the ability of different areas of the world to produce a similar unit of product to that of the USGC (which has been set up as the basis for comparison). So, a 1.0 PF means that the labor force in that area can build/work/produce at the same pace as the US guideline basis. If a particular labor force takes twice as long to produce a given unit of product, that labor force would be described as having a productivity

factor of 2.0. With all that in mind, some typical productivity factors for the same areas listed above might run in the range of:

- USGC: 1.0–1.2 (note: we in the US no longer even meet our own basis of comparison)

- Europe: 1.2–1.5

- Middle East: 1.6–2.0

- Africa: 2.0–5.0

- SE Asia: 3.5–4.0

- Far East: 1.4–2.5

- Australia: 1.1–1.2

So, very simply, if you could identify the number of stick-built work-hours associated with a project, then assuming some low/medium/high module percentages, you can calculate a very high-level potential cost delta between the conventional stick-built approach and a modular approach.

EXAMPLE

(500,000 craft direct work-hours) x (45% modularization or removal of slightly less than ½ direct work-hours from site) x (delta AIWR) x (delta PF)

Site cost to build = (500,000 work-hour project) x (0.45 removed from site) x ($115 for US) x (1.0 PF) = $25.875M

Fabrication yard cost to build = (same 500,000 work-hours) x (.45) x ($30 for Far East fab yard) x (2.0 PF of fab yard) = $13.5M

A potential savings = $25.875M – $13.5M = $12.375M

Note: at the fab yard, the 500,000 x 0.45 has doubled in terms of fab yard work-hours, now requiring 1,000,000 work-hours. But, each fab yard work-hour is only $30 versus the $115 at site; hence the savings noted above.

10.6.1.5 Module versus Stick-Built Comparison

Of course, there is a lot more that goes into such an analysis, and this is where the sophistication of the company module analysis tool comes into play. To have a truly realistic analysis, as noted in the description of the CII costing tool in Chapter 5, the following factors should be considered and included if possible:

- Module type (each has a more or less unique ratio in terms of typical weight/cu. ft as well as weight percentage in terms of steel/piping/equipment/E&I&C/coatings/etc.)

- Module yard (location/craft rate/AIWR/etc.)

- Bulk material differences (specifically concrete and steel)

- Freight and transportation

- Construction management team (CMT) at fab yard (an extra delta cost)

- Module-specific costs (setting/hook-up/carryover work/ scaffolding).

A basis for module cost must be identified. In some tools, it is based on the volumetric size. In others, it is the deck square footage. There are trends for both because if you think about it, as the module gets bigger, it has more deck space, and gets heavier. The trick is to determine an accurate volumetric or areal set of factors based on some historical basis of the typical module types one will be using.

The trick to setting up a good but simple comparison tool is to limit the comparisons to only those major items that are expected to change between the stick-built scope and the module scope. By analyzing only these differences, you simplify the comparison by forcing all the other factors to be equal and therefore canceling each other out of the final comparison. As a result, the suggested analysis reduces to an "easy" cost comparison—the difference between the module cost and the stick-built cost for identical scopes of work. Note that this type of comparison will not get you a total project cost. It will only provide a comparison of the portion of the project scope that was stick-built and has now turned to modular. (And, within this comparison, it will only be on the items that are materially different: the labor costs, some of the material costs like structural steel and concrete, and any special costs, like shipping, setting, and hook-up.)

What Is Similar So, for setting up this type of comparison, the following simplifying assumptions may want to be made:

- **Equipment** is identical for both the stick-built and module cases.

- **Piping** is also assumed to be identical for both cases. This will probably force the cost savings for modular to be a bit conservative since the module case will typically have less overall piping in terms of length. However, sometimes this shorter length is complicated by having more fittings and bends (to remain configured within the module); thus, the suggestion to assume the piping is equal is probably not that far off.

- **E&I&C** is also assumed to be identical. This is probably not always correct, and configurations change. But this cost is small when compared to the equipment, piping, and structural steel cost.

What Is Different
- **Structural steel**, however, is NOT identical. This significant difference needs to be considered and accounted for. The structural steel on a module will almost always be more than its stick-built equivalent. To help one understand why this increase is required, consider the configuration of a stick-built process area. Equipment is set on foundations, and piping is supported around it. For a module, there is a need to create a "bottom" for that module that will be used to support the equipment that was typically set on a separate concrete foundation on the ground in the stick-built scenario. This module "bottom" must also support the weight of the entire module frame and equipment as it ships on the ocean-going vessels, provide connection integrity via grillage to the ship's deck, and provide a flat base to sit on top of the trailers or SPMTs that will be used to move it to the vessel and from the vessel to the final setting location. Thus, the module must have at least one extra level or layer of structural support steel.

- **Concrete volumes** will be different. This is a bit more subtle than the structural steel case but needs to be addressed. Because the module is made up of more steel tonnage, it must be supported by additional concrete, either in terms of more piling or more spread footers. This may be further defined in terms of cost/cubic yard of piling or spread footing. It may even offset some of the higher costs associated with finishing above-grade piers and other concrete supports. But, it is a difference, and an attempt should be made to capture the differences.

Module Size and Weight Correlations Depending on the size and type of module being used, there are direct correlations between the size and the delta weight difference over the similar stick-built scope. It is not an exact relationship because the experience of the equipment and piping module layout person has a lot to do with how "big" or "little" they will make a module when trying to fit this set of equipment into the confines of the structural steel envelope of the module. However, different but equally competent piping layout designers should be reasonably consistent in developing the module size, so the size/weight correlations should not be too far off when comparing one module at one project with a similarly sized one at another project, even if developed by different layout designers.

As an alternative to hiring a competent and experienced layout designer, there are industry estimating programs that will take the equipment list with the proper level of input details and develop the equipment size and corresponding connection space between it and other adjacent equipment, thus providing a sort of volumetric building block for each piece of equipment. By taking all the equipment developed this way, this set of volumetric building blocks can essentially be assembled and stacked or "pushed" together to develop an idea of the overall total module volume. One such program is ACCE (Aspen Capital Cost Estimator™). Of course, the big caution when using this as well as any other estimating tool is the universal tenet—GIGO, garbage in, garbage out. Rely on such tools only to the extent of the data used to populate them.

Other Non-Correlatable Costs   Then there are wildly differing costs depending on the project site that probably should not be included as part of the standard module cost calculation tool but should be estimated and included separately. Examples include the cost of the material offloading facility (MOF) as well as the heavy haul transport road.

Freight   Finally, we suggest that whatever tool you use or develop, that you go ahead and make the simplifying assumption that the total material and equipment freight costs are similar between stick-built and module options (with the exception of the one true delta cost: the final shipping cost of the completed module to site).

With a truly worldwide procurement organization, it is too challenging to try to determine deltas for shipping to a module yard versus shipping directly to the site. So, we suggest that any comparing program keeps it simple and assumes that everything equals out in the end. Think about it, with much of the equipment being fabricated overseas, individual shipments from the vendor to the module fab yard may be much shorter and faster than shipments to the site. In some cases, the module fabrication yard is its own equipment vendor, having internal fabrication capabilities to actually fabricate this equipment on site.

But, as mentioned, the final completed module transport from the fab yard to the project site should be considered as a true delta cost associated with the modularization effort.

10.6.1.6 Running the Various Module Scenarios

Based on the type of analysis tool you have available or the assumptions you have incorporated into the one you are building, begin evaluating the project based on module size options. This is not exactly straightforward, so the following suggestions on trends may be helpful when trying to develop meaningful differentiators between the various sizes of modules.

Develop some module size standards that are uniquely different enough to provide measurable differing metrics that can be assigned to each size standard. We suggest the following ranges as an initial guide:

- Truckable—must meet most State highway clearance requirements.

- Oversize or permit load—extra wide and tall.

- Small equipment (PAU)—typical skid mounted self-contained type equipment, moved by special multiple axled trailers (e.g., Cozad or Perimeter frame).

- Medium equipment (PAU)—larger, containing several levels of equipment and requiring multiple lines of SPMT trailers.

- Large equipment (PAU)—largest, requiring even more lines of trailers and typically fully containing a sub-process system.

- Small pipe rack (PAR)—typical one or two levels—no equipment.

- Medium pipe rack (PAR)—several levels—no equipment.

- Large pipe rack (PAR)—several levels and complex, potentially with fin-fan coolers and other equipment incorporated.

Each should be sized to reflect the full range of what your company expects to see and work with. Be careful not to constrain boundaries with little variation between the options. They need to be robust enough to show significant differences in the percentage steel increase, fabrication timing, shipping costs, etc.

Try to understand and evaluate the compromises that occur:

- Smaller modules use less steel but may not contain complete systems, so they require more connections within the module as well as more connections between modules when they arrive at the site. But they can be built faster and shipped on smaller vessels or even trucked to site.

- Large modules use more steel but can incorporate complete process systems, allowing for more extensive piping, E&I&C hook-ups, pre-commissioning, and testing at the fab yard. However, they take longer to build and require more interaction with procurement and vendors. They are required to be shipped on larger vessels that require earlier contractual commitments, which can become schedule critical if the vessel arrives on time, but the module is running late and is shipped incomplete. They require special MOFs, heavy haul routes, and extra allowances on the site in terms of heavy haul road width.

- All of the above are examples of cost deltas that should be identified and compared in order to determine optimal initial module size.

And, if you have not noticed by now, these trade-offs impact the schedule, the number of craft at the site, and the duration. They also feed back into the scheduling and sequencing of the engineering and procurement activities. So, this comparison effort should not be performed in a vacuum.

10.6.2 Step #2 The Module Layout

Once you have determined the optimal size of the modules for your job (either based on true analysis as discussed above or simply because of shipping limitations to the site), the next effort is turning a typical two-dimensional stick-built layout into a three-dimensional set of modules. This is where the experienced piping layout person is your key to success. This is also why it is so important that when laying out a module job, you do not succumb to the temptation to just "draw boxes" around

the equipment already in that stick-built configuration. Note, we provided examples in Chapter 4 of problems you can get into by simply "boxing" stick-built areas. The module layout was also discussed in some detail in Chapter 5, but a couple of additional points should be made with respect to performing these individual module layouts.

10.6.2.1 Process and Piping

Work closely with the process and licensor engineering teams. They always have some "sacred cows" in terms of equipment arrangement or piping layout that are not to be messed with. Determine what is "sacred" about their process layout in terms of equipment, spatial relationships, configurations, etc., and strive to understand what makes these configurations special. If modifications are not to be made, strive to keep them that way. Remember, we are not offshore, and because of that, we have the capability of genuinely keeping these configurations identical—if necessary.

For all other equipment, determine the best fit within the module. For example, if a vessel's contents must gravity flow into another vessel, consider setting this vessel on an elevated or second deck. If air-cooled equipment needs to be in an open area, consider setting them on top of the module. The goal is to use the third dimension (vertical) to your advantage (and continue to remind the process, licensor, and equipment designers of this increased flexibility and opportunity).

10.6.2.2 Piping Flow Diagrams (PFDs)

Use the piping flow diagrams (PFDs) as a guide for module content. The PFDs are high-level simplified explanations of the process. We may have oversimplified the process engineering's work, but the PFD is typically drawn to show a more or less complete sub-process on a single page. So, look at how the PFDs are divided up, this may give you a clue as to how one might divide the proposed plant into separate modules. And, because each PFD page represents a more or less self-sufficient sub-process, a module encompassing all that equipment on that PFD page will also encompass all the E&I&C work needed to make it run. By designing in this manner, you will place the fabrication yard in a good position to be able to complete the module and perform a significant amount of pre-commissioning and testing of the corresponding E&I&C work prior to shipping to the site.

10.6.2.3 Operations and Maintenance (O&M)

Remember to involve operations and maintenance (O&M). We like to think of the O&M team as the "ultimate and final customer." As you develop the module configuration, make sure that the O&M team who will actually be working daily on and in these modules are taken care of. Make sure they understand what the modules will look like and what they will be left to work with.

For example, O&M needs to understand that the first level may be as high as 4' to 10' above grade on a large land-based module. THIS CAN BE A BIG DEAL! Do not underestimate the power of influence that O&M can bring to bear on the design, after all, they are the end-user. Make sure that the design not only covers their needs but does not cause heartache due to a failure to communicate.

On the positive side, involving them early can provide great insight into operations that may result in potential savings or a simpler way to configure. For example, discussions with the O&M team knowledgeable in the day-to-day running and maintenance of the plant may reveal opportunities to eliminate or reduce permanent access to places where daily, weekly, or monthly visits are not required.

In other cases, there may be a need for extra room to either charge a vessel or remove its contents. Such slight shifts (try to say that three times quickly) may provide a much easier, faster, and safer way to perform periodic operations such as routine catalyst change-outs or vessel work.

10.6.2.4 The Module Layout Designer

A key to this effort is a competent (experienced) piping layout person. It is the module layout designer's job to take the typical two-dimensional (2D) stick-built process plot plan and convert it to the three-dimensional (3D) modular layout. The attributes of such a person are:

- the ability to understand the entire process in terms of pressure, pumps, flows, etc. (e.g., what needs head and what gravity flows);

- the ability to sit with the process engineer and work through options on equipment locations in three dimensions (and critically propose options on equipment layout that provide a well-shaped module). For example, pumps and compressors now have the option of being set on levels above ground level. Drums can now gravity feed to lower levels.

- a good feel for the spacing requirements between equipment pieces that are identified on these PFDs to be placed in the module:

  - based on an understanding and preconceived knowledge of how the piping and other "stuff" (not currently shown on the PFD) must eventually fit;

  - understanding the maintenance and accessibility requirements of the equipment pieces;

  - understanding E&I&C needs (and these accessibility requirements);

  - providing for all the safety requirements—walkways/access ways/exits/stairs/ladders.

- a vision of what the completed module should look like in terms of accurate dimensioning without running much, if any, piping.

Proper module sizing for the equipment is critical. Too much space between equipment will create a larger than necessary module size (and corresponding module cost). The opposite is just as bad, requiring a later rework to "expand" one or more modules due to congestion as the piping eventually is laid out. As this effort is developing, frequent reviews and updates should be provided to the entire team, again because of the interrelationship of all the players besides the obvious engineering interactions.

This brings us to the next subject, long lead equipment, before we get too far into the details of the module and its contents.

10.6.2.5 Long Lead Equipment (a Digression)

This is probably a good time to discuss the issue of long lead (LL) equipment. LL equipment is typically defined as all equipment that takes 12–18 months (or some other duration requiring advance commitments by the owner) to procure in order to meet schedule. This is usually a straightforward list to develop on a stick-built project.

However, for the modular job, there may be other equipment that may fall into the LL equipment category due to the early module fab yard timing requirements. These requirements include the need to be able to receive complete structural steel design drawings to the level of detail that these can be used to actually begin fabrication of the steel: this level of detail is typically issued for construction (IFC) drawings. This means that all the equipment in the modules will need to have been designed, sent out for inquiry, reviewed, accepted, awarded, and early engineering requirements, such as layout and configuration, be finalized so the structural steel can be detailed to provide accurate support. Beyond this, the actual fabrication of this equipment must be completed soon enough to be shipped to the module fabricator for installation on their appropriate module levels at the appropriate times. This becomes critical for equipment on the lowest levels of an equipment module as well as for any high alloy coolers required to be set on central pipe racks to be shipped early to the site.

Such a list is not obvious without either some preliminary module layout work, as performed in Step #2, where the proposed module has now been configured and the major equipment identified or at least some assumptions made on what might be modularized. Such an assumed list can "easily" be developed early in the project assessment phase based on preliminary options analyses of the number and type of modules being considered. This list is typically developed early in the project planning on any project. To be safe, one should assume the max module case to see what equipment might fall into the group with timing constraints based on the proposed engineering/procurement cycle. Whichever way this is worked, the goal is to identify equipment that must have some amount of engineering purchased prior to FID in order to meet schedule and then perform an analysis of benefits to either "leave on" versus "take off" the module depending on the project drivers.

There can also be an issue for any electrical or instrumentation buildings (or major equipment) that must be incorporated into the bottom areas of large complex equipment modules that are required at the fab yard early for installation. Fortunately, much of the equipment in a module is not overly complex or time-consuming to fabricate, and it will allow for some flexibility in terms of a later than optimal purchasing effort.

10.6.2.6 The Three-Month Rule

One of the earliest scheduling efforts with respect to modules is a sanity check on the delivery timing of every piece of equipment destined to the fabrication yard. (The sanity check is to confirm that all will be at the fab yard within 3 months of the start of the structural steel fabrication efforts—the first steel cut.)

You probably won't find this "3-month" delivery rule in many module commentaries. You may even find people who will argue with you about its validity. It is not a hard and fast rule that cannot be broken or bent and still meet schedule. It is more of a "guide" for those of us who have struggled with meeting a fab yard's schedule. If followed religiously, the 3-month rule will cover a "multitude of sins" when it comes to procurement issues and meeting the fab yard's needs.

Why 3 months? Because the average time in a fab yard (on a large module) from the first cut of steel to the point where there is enough steel to start assembly of a large deck is about 3 months. If all the equipment is on site within this 3-month window, the fab yard has complete flexibility in setting whatever/whenever/wherever in its efforts to assemble in the most efficient manner it can.

10.6.2.7 Equipment with Delivery Problems

As inevitably as fall follows spring and summer, equipment will be delayed. There will always be the challenge of getting everything to the fab yard on time. So, what happens in the initial review of equipment deliveries that you find a piece of equipment will not make the required delivery date with the module fab yard? There are essentially three options (if the original decision to send it to the module fab yard was based on sound reasoning):

- Remove the piece of equipment and ship it directly to the project site for installation (**easiest decision**—but potentially the worst option).

- Work with the fab yard on options to install it late (**better decision**—because it was sent to the fab yard in the first place due to the importance of installation off site).

- Design and purchase it early enough (maybe even get some engineering funding prior to FID) so that critical engineering deliverables and later the equipment itself can be purchased to support the fab yard delivery schedule (**best decision**—as this gets the equipment off site and to the module fab yard in time).

The final decision depends on the project drivers and options available to the project, keeping in mind that the final decision needs to be the one that will have the least impact on cost, schedule, and the site construction team, not necessarily in that order.

The next issues to be faced are the subsequent late equipment deliveries due to any one of a thousand delays no one can anticipate. The best solution for mitigation of these impacts is to closely monitor the progress of equipment fabrication and identify any potential delays as soon as possible. By doing so, you leave yourself with two options:

- Work with the fabrication yard on alternative installation schedules for a late delivery.

- Expedite the equipment vendor.

By catching delays early, there may be workarounds to accommodate the late delivery.

10.6.2.8 Module Equipment Exceptions

Prior to this review of the equipment that goes into a module, it is assumed that the following initial efforts at optimizing the module scope and content have been made. Some early analysis and decisions on what *might not* go into a module can be straightforward, for example:

- **Large rotating equipment.** Incorporating very large rotating equipment into an equipment module will require dynamic analyses. Depending on size and type, this may require complex vibration and/or harmonic analyses that may not be a risk that the project wants to accept.

- **Large, heavy, or tall towers**. Incorporating very large and heavy towers into a module may not make good project sense if the connections to/from the tower are relatively simple in terms of final tie-in to the rest of the plant (and the number of craft work-hours transferred from the site is minimal). Performing such work at the site may be more cost- and schedule-effective than designing the tower for the additional transportation loading and designing the module with all the additional structural supports required to withstand the one-time stresses of grueling sea transport in heavy seas.

- **Tanks and other large or voluminous vessels.** Building a structural "box" to incorporate a thin-walled vessel or tank may not be the most efficient use of project resources. Often, it does not make sense to modularize a lot of "air."

Now, back to the 5-step module analysis.

10.6.3 Step #3 Expand and Accept

As the modules are defined in Step #2 and sorted out on the plot plan, there will inevitably be concerns with access between, around, and or through the module complex requiring some sort of rearrangement.

Despite how thorough your early reviews were and how inclusive the attendee list was, it may be necessary to rearrange:

- equipment and piping within the module;

- equipment and piping between the modules and the stick-built portion of the plant;

- the modules themselves.

This may be driven by details, including maintenance/safety/accessibility concerns that were previously not identified or discussed during the initial module configuration discussions. It could be due to fire/blast concerns that may be newly identified after these detailed analyses were performed. This may be due to the need for additional spacing for some of the underground utilities, the large motor control centers (MCCs), or other units. Whatever the reason, after the initial module arrangement has been developed, it is good to have a formal review with all stakeholders to make sure all concerns have been covered on their sizing and configuration.

While the essence of this third step did not take long to explain, this step can cause major re-work if early reviews are not complete, and as a result, significant changes are identified and required. In one such example, the owner/client failed to identify the need to initially and then periodically "bench test" all control valves that had been installed in the modules, some of which weighed several tons. Because of the maturity of the module design, this required major re-work to several of the process modules, including the addition of structural steel for monorails, davits, and other overhead methods of access to get to each control valve as well as redesigned and additionally reinforced walkways and access ways to support the larger capacity wheeled carts to take the valves to the edge of the module for retrieval or lowering to the ground from newly designed extended platforms.

Such last-minute design re-work can be minimized by ensuring that all project stakeholders have been "required" to attend the various 3D model reviews.

10.6.4 Step #4 Schedule

Scheduling is critical in any job, but because of the need for so many deliverables in the middle of the project during engineering and procurement to support the fabrication yard, it becomes very important that the schedule supporting this fab yard (including all predecessors that snake back into these groups) have been identified and their interrelationships confirmed.

So, what does this mean, and how do you approach such an effort?

10.6.4.1 Scheduler

First, you need a scheduler. This is not just someone that is good at Primavera P6 or other software. This is someone who truly understands how the project groups interact and can deftly walk anyone through the various parts, including primary and secondary critical paths. This person needs to understand the differences in approach between the stick-built and modular projects.

10.6.4.2 The Stick-Built Schedule versus the Modular Schedule

A brief comparison follows of the stick-built schedule versus the modular schedule.

The Stick-Built Schedule Take the standard scheduling effort and look at what its goal is. Typically, its goal is to get everything to the construction site in time for field erection and assembly. Since everything will be assembled at the construction site, the main focus becomes ensuring the site path of construction is supported. But, in many cases, this site path of construction is somewhat flexible and offers alternative paths should parts be slightly delayed. Because of this, such a schedule, that allows minor delays on some structural steel, equipment, and piping due to poor scheduling and inaccurate ties, may be able to be tolerated because of the inherent flexibility of the site path of construction, allowing the project site team to shift priorities (at least to a certain extent).

Of course, we are not saying anything can be shipped at any time to the project site, and the construction team will be able to handle it (even if there still exist some construction managers who do boast this ability). But we are saying that flexibility typically exists to accommodate some of these incongruities of scheduling and delivery.

The Modular Schedule Now, how does that compare to the modular project? As mentioned in Chapter 8, for the modular project, there will be a large portion of the engineering and procurement planning efforts as well as a large amount of structural steel, equipment, and piping that must now be accelerated in terms of detailed design to meet the intermediate goals of getting a completed design to the module fabrication yard. While the overall level 1 schedule will still show the major deliverables in terms of steel and piping design in the relatively same month of delivery, the early planning efforts and details required to get these to an IFC quality for the modular job are much more complex.

For example, for structural steel, instead of designing for supporting major equipment on concrete foundations, there must now be enough early equipment design to be able to design the structural support steel for this equipment. For piping, the stress analysis must be completed to the point that it can be incorporated into the structural steel design rather than allowing for a subsequent design follow-up effort that may accommodate stress needs outside the module.

Then, since some of these large modules may take as long as 12–18 months to complete, there will be a higher priority on making sure deliveries to the fabrication yard are timely. Late designs or changes to piping and equipment will have a more significant negative impact on the module fabrication effort and the module schedule. And, unfortunately, because the module is a single composite piece of design work, the entire module (and all of its pieces) need to be designed to the point where the overall design of the module can be complete. To delay the module build in the middle of its assembly to wait on late equipment or to modify the structural steel on a completed module design in order to accommodate a size change essentially shuts the entire manufacturing process down. The fab yard is like a car assembly line—the sequence of work is specific and crucial, for example, there is only one point in the car assembly process where the engine is installed. This is the same with the module build—there is a set sequence for the module build—later efforts depend on earlier completed work.

Compare that to the stick-built project, where parts of an area's erection efforts may be able to be delayed while the missing piece is shipped or the design is completed.

- Is it an inconvenience? Yes.

- Does it have the potential to shut the entire construction site down like it did on the module example? Probably not.

A lot of words for a simple concept—the module project must support the fabrication yard schedule.

Backward Schedule Pass   How do you do this? There are many ways to explain how this is done, but essentially it is developing a very complete backward pass of the schedule. Working with the end in mind—that being the final completed project plant process—in operation and running, the schedule is reviewed in terms of each set of predecessors required to meet the project schedule. But, the key to this backward pass, at whatever level of detail it is developed, is to confirm the scheduled durations are indeed realistic and not based on a best-case, perfect execution, or the one-in-a-thousand chance that everything will work exactly as planned.

The point we do want to get across with respect to this backward pass is that it is meant to define pinch points in the schedule. Points where the sequence forward does not have sufficient time to complete all the predecessor tasks properly.

AIR COOLER PROCUREMENT EXAMPLE

For example, take the seemingly simple effort of scheduling some high alloy fin-fan coolers for setting on top of the main pipe rack to be modularized and shipped to the site (see Figure 10.1). If we walk through the schedule in the typical forward manner, we see no special concerns with respect to the air cooler delivery and installation on the pipe racks. The schedule in Figure 10.1 seems reasonable in terms of durations for the pipe rack module material purchase, piping installation, and cooler setting.

It is only when we make the backward pass on this schedule that includes adding the durations for the bidding and purchase of these air coolers, that we see a completely different situation (see Figure 10.2).

It is discovered that the lead time for the actual fabrication of these specific coolers is so long that in order to have the equipment at the module pipe rack fabrication yard in time for its installation on the pipe racks (month 10½), the initial procurement process must be initiated prior to FID by at least 6 months. This is the only way to allocate enough time to complete all the preliminary purchasing activities (in terms of three bids and a buy) and conditionally award this package to the successful bidder in time to permit engineering and the vendor to work through the detailed design requirements by the time FID arrives. The reason for this extra early effort is the duration on the fabrication of these high alloy coolers:10 months from award.

Of course, this requires special contracting efforts, including conditional acceptance and limitation of initial scope that will allow award commitment and preliminary engineering to be performed prior to FID and prior to actual award of the fabrication.

At FID, when the project has been funded and the EPC contract awarded, the vendor can be directed to start the fabrication efforts, which will consume all of the time available to get the completed air coolers to the fabrication yard for a shipment and setting by month 14.

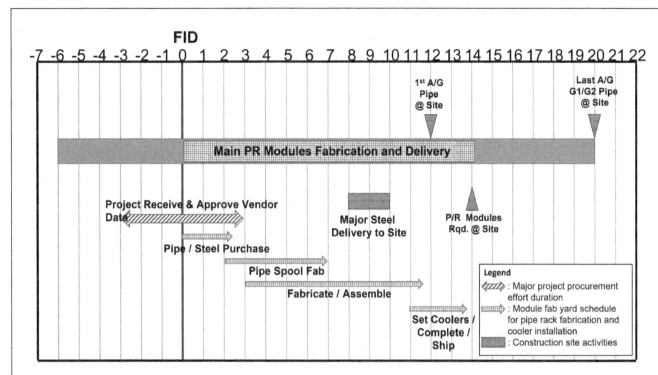

Figure 10.1 Typical schedule forward pass.

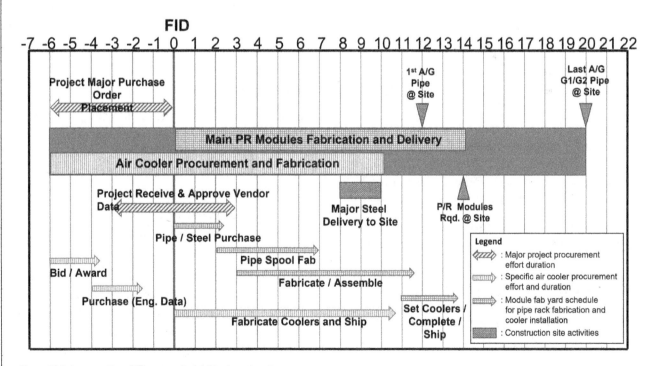

Figure 10.2 Incorporation of "Reverse schedule" backward pass.

So, what sort of typical "Reverse" schedule chains need to be considered? Anything that has the potential to delay the delivery of information and/or materials to the module fabrication yard.

The following examples show some typical activity chains in reverse:

- **Overall project (through module fabrication yard):** O&M/Start-up/Commissioning/Pre-commissioning/Path of Construction/Module set sequence/Module delivery/Module completion/Module assembly & fabrication/Pre-assembly/Equipment delivery/Bulk material delivery/MTOs/Procurement/Engineering design/Process design/Project Execution Plan/Design Basis, etc.

- **Overall project (through LL equipment):** O&M/Start-up/Commissioning/Pre-commissioning/Path of Construction/Long Lead equipment setting/Foundations/Piling/Site Prep/Mob to Site/Construction schedule/Rigging schedule/Underground schedule, etc.

- **Overall project (through site development):** O&M/Start-up/Commissioning/Pre-commissioning/Path of Construction/Long Lead equipment setting/Module delivery/Site equipment delivery/Foundations/Piling/Underground/Foundation design/Equipment details/Site Prep/Mob to Site/Construction plan, etc.

Each of these steps may have additional internal backward pass requirements, for example:

- **Engineering (materials related):** I&C design/Electrical design/Piping spool development/2nd pipe MTO/Steel IFC drawings/1st pipe MTO/initial steel MTOs/Foundation and Underground IFCs/Piling IFCs/Site Prep IFCs/Civil Design/Plot Plan layout/Project Execution Plan, etc.

- **Engineering (model-related):** Piping isos/3D Model Build/Stage III Model Review/Vendor Data/Stage II Model Review/Prelim Equip Info/Safety Reviews/P&IDs/PFDs.

- **Procurement:** Shipping/Equip Manufacture/Award/Bid Evaluation/Request for Quote (RFQ)/Solicitation of Interest (SOI)/Engineering Technical Requisition/Engineering Design/P&IDs/PFDs.

10.6.5 Step #5 Cost

This step is typically worked in multiple stages, from the initial +/- 40% to 50% accuracy to the final +/- 10% effort

required on some projects to get management approval from both sides—the owner in terms of project approval and the contractor in terms of execution approval.

For our final step in the 5-step module estimating process, the first effort is to agree on the cost basis, the cost inputs, and the necessary accuracy. Accuracy is of utmost importance as it drives how much time the project team is going to need to expend to develop engineering and design to the level to get to that accuracy. Typically, when it comes to a modular execution plan, the comparison number that many want is the difference in cost (as in project savings) for following a modular rather than a stick-built execution plan. So, if you have never performed such an analysis and do not have metrics on the comparison, how should you proceed?

The first step proposed in the 5-step plan provides a method of making such a comparison. For all subsequent cost analysis efforts, this can be followed again, but with more details available. The goal in all of these iterations is to identify the major contributors to both the additional costs associated with the module decision as well as the additional savings resulting from this same decision. We have identified them in Step #1, but a high-level re-cap is worthwhile.

10.6.5.1 Module Savings

- Direct—Site labor. Defined by the comparison of economic productivity ratios—Site versus fab yard, which is the combination of AIWR and productivity factors. Check the details of the economic productivity ratios in Chapter 5.

- Direct—Time to market. Module schedule reduction in some projects equates to improved NPV and reduced TIC.

- Other indirect or soft savings. Defined as the benefits to the project due to moving this portion of the project to a place where it can be built in better weather conditions, at grade, with an experienced stable workforce, thus reducing impact at site, in terms of temporary construction laydown area, traffic, noise, congestion, etc.

- The ability to pre-commission a higher percentage of this equipment and systems prior to arrival on site.

10.6.5.2 Module Costs

Engineering and materials related to the module configuration, shipping, and final site placement, including the following:

- extra engineering efforts related to the module configuration and coordination;

- extra structural steel;

- additional construction management team (CMT) at the fab yard;

- inspection/QA/QC/preventative maintenance—on equipment that now must sit longer at the fab yard (installed in modules) prior to being commissioned;

- the entire shipping effort of the completed module;

- the module movement from receiving MOF to the site, heavy haul route development, module final setting, and hook-up.

The goal in all of this is to identify in sufficient detail, all major cost impacts (both pluses and minuses), and then get everyone in the project team that might be familiar with costing them talking about the actual delta costs. This is best done in a group session rather than individually, as there is more synergy with several people thinking about ramifications rather than only one or two.

This effort is very similar in format to some of the more formal risk reviews that are set up throughout the project duration.

10.6.6 Other Activities

At this point in this example "real-life" module analysis, all the major activities of the module "shaping" have been addressed (or should have been addressed), and along with them, many of the other activities identified and analyzed that are important in the development and support of the module effort, including:

- Fabrication yard—evaluation, selection, and award.

- Module logistics—really an extension of the equipment logistics all projects are familiar with, but on a grander scale in terms of vessel and route selection, as these have an impact on the module design for shipping accelerations.

- MOF and heavy haul—routes and coordination.

- Path of construction—to site to support a specific module setting sequence.

10.7 Concerns to Watch Out For (Lessons Learned)

This is one of those subjects that can easily turn into another stream of consciousness dump from the experiences of the two authors. Where do we go first to provide some cautionary comments and discussion, and how do we even try to link these comments and observations into some logical stepwise progression for the reader? This could be an entirely separate chapter or even an entire book.

They are challenges that seem to come up throughout the project execution effort. Since they seem to be able to be grouped into a few summary topics, a quick run-through in terms of approximate chronological order or project sequence may be the best approach. Many of the finer points have been identified in Chapter 8, Modularization, but they deserve a second consideration because they offer extra and typically unwanted project complexity if not addressed and resolved early:

- **Structural members and connections (otherwise known as bolted versus welded modules):** Consider pros and cons, local labor preferences, fab yard capabilities, proximity to galvanizing, methods of assembly and erection, design and limitations, protective coating requirements, and durations.

- **Module yard selection:** From due diligence to properly identifying fab yards that can efficiently complete the proposed scope. Match modules to fab yard expertise and capacity, proximity to the site, CMT requirements.

- **Static and dynamic analyses:** Should be considered on all large modules, impacts both piping and equipment design, impacts how the module is prepped for shipment in terms of tie-downs and temporary bracing.

- **Grillage and sea-fastening:** Try to develop a simple design, easy to install and easy to remove, flexible for module location on deck and reusable (if logistics allow).

- **Logistics and transportation:** First analysis of a module job, contract an expert, look at all options, and you get what you pay for, so go with someone who knows the area, has the equipment options, and the experience in option evaluation.

- **Pre-commissioning and commissioning:** Make sure planning is complete, approval process is understood and accepted by all, paperwork is complete and gets to site in time, meaning of approval signature authority is known and understood by all, make team stand by decisions made in terms of sign-off.

- **Setting and hook-up:** Evaluate single weld hook-up concepts and scope, make sure design takes into account temperature and transport configurations, make sure path of construction allows removal of SPMTs after setting, consider long-term corrosion issues around foundation connections at grade.

10.8 The Inevitable Question

Finally, it would not be a complete "real-life" scenario if we do not address how to handle the option offered by many well-meaning owner/client organizations and even some EPC contractors: a late request for a module study option.

10.8.1 The Late Requested Module Study

This request is typically the result of the well-meaning application of some gross misinformation from various inaccurate sources on module planning and its subsequent implementation. The wrong conclusion resulting from this misinformation was that the module effort decision for analysis could be delayed until FEED or even into early EPC.

Of course, the well-intentioned contractor's management will accept the challenge and immediately come to you—the up-and-coming company module expert—for an answer and acceptable resolution. Telling your boss and the client that they have completely missed the boat in terms of modular opportunity is not in your company's best interest and could be considered a career-limiting move for you, the up-and-coming module expert. In addition, there actually are some efforts that may result in some benefits, even at this late juncture in the project development.

So, we suggest the following option as an example of how one might respond to this challenge. Obviously, this is not a one-size-fits-all response, but it will identify some options that may be pursued with respect to such a late request (without telling your boss and the owner/client they have collectively lost their minds).

10.8.1.1 Why Is the Late Request Such a Problem?

First, let's understand why this is such a problem in terms of actually developing a module solution. If such a scenario is initially brought up this late in the project development, obtaining any modular solution, much less an optimal one, becomes difficult due to work already performed and decisions already made.

Consider the many module impacting decisions already made in the previous project phases. The site has been identified, the project equipment selected and arranged in a selected (stick-built) plot plan layout, the route to this site has been examined only in terms of supporting individual equipment shipments. Open water access may not have been considered, and as such is probably not available. If it is, the shoreline is probably not equipped with an MOF adequate to support offloading of anything more than the previously mentioned equipment.

With respect to a project team to offer effective and timely consultation on the subject of modularization—there is none. With no module team identified or in place, there can be no consistent, uniform direction or support of any new modular goals or new project direction.

Plus, there may be little interest in developing any module alternative among most of the contractor's management as well as probably all of the rank-and-file project members (due to potential re-work requirements with no accompanying relaxation of project due dates by the owner/client).

Any potential modular solution now being examined has potential ramifications to change everything designed and purchased to date, which could be significant. Changes to the current plot plan will impact everything civil/structural has done in terms of site design, grading, and undergrounds. This would bleed into the piping and equipment layout efforts to date, where the modular solution would require a re-work of any piping and equipment layout work. And, even if all that could be magically changed with the snap of the fingers, there would still be the inevitable scheduling and early engineering design issues for most of the equipment that may now be destined for a module.

10.8.1.2 Where Does a Project Start with Such a Late Request?

Go back to the basic steps for developing the modular solution that should have been identified and worked through in the opportunity framing, assessment, and selection phases of the project. The first step would be to commission a heavy haul route survey to be worked immediately to determine the maximum shipping envelope to the site. (This is a must, as it does the project no good if any of the resulting modular options are too big or heavy to be moved to the site.) Then, specifically, as suggested in Chapter 5, follow up with initially identifying the project drivers that were the reason for bringing this modular alternative up, even if it was so late in the project development and what the potential benefits that might still be available as well as the efforts involved to achieve these benefits.

What-if assessments should be performed with key members of the entire project team: the owner/client and contractor management and all stakeholders who might be affected by this change in philosophy.

Each project driver should be discussed in terms of what the module alternative might provide. If one driver is cost, determine from a project cost standpoint, the potential maximum $ savings that could be realized in terms of the ideal solution of maximum modularization of the facilities and the biggest differences in cost and productivities between site labor and fabrication yard capabilities. If it is schedule, take the same approach in terms of best schedule savings by maximizing the more efficient building off the site.

Then look at where the current design effort is. How much of the work to date would need to be scrapped and re-worked to facilitate the maximum module scenario? How much can be salvaged and, with some modifications, re-worked into the modular approach? What are the drawbacks?

The most likely initial knee-jerk reaction received from both sides in terms of the resulting module effort is that the cost savings would probably be minimal, and there is no way to recover on the resulting schedule delay. This is where some good pre-planned what-if scenarios and a competent scheduler come in handy. If the project management is honest with themselves in terms of the actual on-site labor cost and productivity, they will probably discover that the project end date (e.g., ready for startup, RFSU) extends farther out than what is currently shown in the schedule. The argument should be made that the maximum modularization of any part of the project will provide a better "schedule surety," and the dates promised by the fab yard will have a higher chance of being met, leaving less to chance with the smaller on-site workforce.

If this is the case, the exercise then becomes an engineering discipline-by-discipline analysis of which efforts can be salvaged and what must be re-worked, and what can be used when moving the maximum scope of work to a module-based effort.

The goal of this analysis would be to identify something significant in terms of meeting one or more key project drivers. Anything less may render the maximum module option a non-starter since such a quick and high-level comparative analysis will have unidentified circumstances that will reduce the initial benefits identified in this analysis.

10.8.2 Less Ambitious Options

So, if the initial analysis is marginal at best, what next? This should not be the end of the road in terms of analysis of module potential. The subsequent analysis should be with a lesser percentage modularization, perhaps leaving out critical equipment and parts of the project plant that make the maximum module alternative a non-starter. For example, analyze a scenario that takes out large rotating equipment, tall, complex, and heavy vessels, and other equipment that is complex or long lead and assume they will be set on site. Work equipment modules around the equipment that is not complex, difficult to design and procure, and can be delivered in time to meet a fabrication yard's schedule. Combine this with an effort to match the resulting equipment module configurations to any limitation identified in terms of the shipping envelope to the site.

If further consideration of the above still results in a marginal benefit due to work already completed or schedule concerns, evaluate the scenario where the very basic elements of the project are examined for potential modularization—the pipe racks and obvious parts of the process, like utilities, that can be packaged as skid units. The unique aspect of a pipe rack is it is simple in design and materials can be procured early. Converting a stick-built pipe rack to a modular pipe rack is also relatively straightforward as the module version may not change much from the original stick-built version. In addition, these pipe racks can be adjusted in terms of overall length and width to accommodate existing road restrictions.

Utility sub-systems should also be considered. They have multiple equipment and complex piping arrangements and have been designed to work as a unit. Because of this, they can easily be packaged and, in some cases, have previously been partially packaged by their vendors. It would not take much effort. Much of the effort to develop a package that could be assembled off site and shipped as a single or common skid could be worked by the equipment vendors, who would be eager to be able to provide a product with more mobility and utility.

10.9 A Couple of Observations from Experience

10.9.1 Underestimation

What will most assuredly happen is that the project will underestimate the benefits of even this late module redesign, overestimate the amount of effort to convert, and also overestimate the construction site labor's ability to meet both cost and schedule goals identified in the project schedule. As a result, the actual modular scope selected will probably be less than what would have been the project optimal scope based on circumstances.

10.9.2 Under-Collaboration

To make such a late change really impactful on a project, there must be a high degree of honesty and openness in terms of the potential benefits to the owner/client in terms of project metrics (i.e., NPV). By the same token, the contractor will need to be willing to share actual costs, schedule issues, and risks in terms of execution with the owner. Both will need to come to an agreement on acceptable contractual modifications where the benefits and risks are appropriately shared: the owner/client in terms of schedule and cost considerations, and the contractor in terms of a reduced downside risk due to the late re-work.

As a parting example, in a project where the FEED deliverables were poorly developed requiring a re-work of the design basis, it was not until both owner/client and contractor met and agreed there was a problem and contract terms were modified to more evenly distribute the risks to both parties, that the project was able to move forward efficiently. While many project team members were not intimately familiar with the contract detail changes, what was immediately evident, was that once this change was implemented, there was a much more open collaborative effort between the major parties. Since any project optimization now benefitted both, project issues were evaluated in terms of impact to the cost/schedule/etc. rather than simply meeting to discuss the failure to meet contract obligations.

10.10 Conclusion

A couple of thoughts on our goals with respect to this chapter:

Identify an approach.

We offered an idea of how to approach the analysis for a project in terms of planning while at the same time trying to communicate that any approach shown is not a "one-size-fits-all." Fortunately, despite the variability on the different approaches and what to do, there are some basics as well as an orderly approach that is somewhat universal.

Provide an example.

So, rather than provide a bunch of lists and grouped ideas on what to look at in terms of the planning for modularization, we felt that a "walk-through" of a "typical" project module analysis might make more sense. Such an approach would be easier to read, more sequentially organized, and perhaps, a bit more entertaining. So, we walked through a step-by-step process, offered some typical push-back scenarios that we have seen, and suggested how to approach some of the more common project variations during the duration of the early planning efforts. As such, we hope that our goal was met. Since this effort is not a be-all to end-all module planning example, we continue to encourage open and frank conversations within the project as well as more personal research outside the project on ways to improve such a planning process.

As we mentioned earlier, planning for a modular job is not rocket science. Most of it is good common sense and utilization of practices and guides that have been around for a long time. But more collaboration is essential in improving such planning efforts, and we encourage making the effort to develop the relationships that improve this throughout the project organization and beyond.

chapter 11

Modularization Application Case Study Exercise

11.1 Oil and Gas (Downstream) LNG Modularization Project Scenario[1]

What follows is a hypothetical case project with activities for the reader to think about in terms of what they have read and understand up to this point about the process of modularization (Figure 11.1).

11.2 General Project Description and Background

Big Oil Ltd has recently awarded a large EPC firm an estimated $10 billion contract for engineering, procurement and construction management (EPCM) of a liquefied natural gas (LNG) project (see Figure 11.1). The project was awarded to this large EPC firm after they successfully navigated through the various earlier phases of the project development, as explained in more detail below.

This large EPC firm will become Big Oil's downstream engineering, procurement, and construction management contractor for this project. They were brought in early by Big Oil Ltd and have successfully maintained a creative

edge with respect to their competition, maintaining a consistent approach to the execution of this large project.

The large EPC firm will be executing and constructing the LNG facility on Opportunity Island, an island on the east coast of Africa, located on the relatively shallow and wide North Kenya Bank and the northernmost part of the Tanzania Zanzibar Archipelago (Figure 11.2). The closest major geographical landmark is Pemba island (a lower portion of the Tanzania Zanzibar Archipelago), located approximately 160 miles (260 kilometers) SSW of Opportunity Island.

The LNG facility will consist of two 5 million tons per annum (MTPA) LNG trains, gas processing and treatment facilities, product storage and offloading, complete off sites, utilities, and accommodations. Timing is critical for completing these facilities as it must coincide with the offshore drilling program and the initial gas well completion efforts.

Figure 11.1 A hypothetical case project. Source: Bim/E+/Getty Images.

---

[1]Disclaimer: This case study is not an actual or proposed project. The location is imaginary—there is no specific Opportunity Island site on the east coast of Africa, so the characteristics assigned to this island are purely hypothetical and are not based on any actual or historical facts. The scenario has identified this specific area only to set a plausible background for the detailed examination of how a project might be progressed in terms of the module potential for a general remote facility and how such a facility might be supported worldwide. Any resemblance to facts about the area is unintentional and purely coincidental.

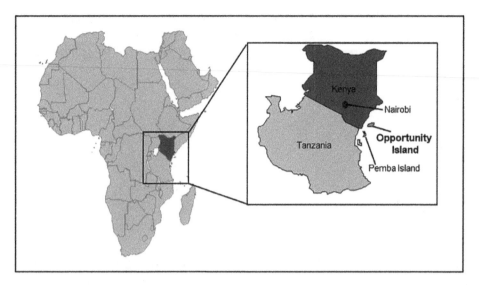

Figure 11.2 Location of the site.

11.3 Additional Project Site/Existing Facilities Information

The following is some additional basic information that the readers may find interesting in terms of background information for use in the analysis and development of the various project phases.

11.3.1 General Information

The proposed project's scope includes the design of the LNG facility, consisting of two 5 MTPA LNG trains and a 200TJ/d domestic gas plant. Big Oil has proposed that the early study efforts, as well as the FEED and subsequent EPC execution, be conducted from their main offices. For the large EPC firms in contention, that would be out of their various operating centers in Houston, Texas, USA.

The successful EPC contractor will eventually set up project offices in Nairobi, Kenya. It is thought that the local construction office and craft worker camp will be set up on either Opportunity Island or the mainland, depending on the decision of where the LNG plant will be built.

The Big Oil downstream oversight team has its project office in Nairobi. This office will lead the project and house the project senior management as well as a number of technical and support service teams there.

The main construction management team (CMT) for the EPC contractor will be housed along with their local craft on site—again, location based on the outcome of the analysis of the various project site options.

11.3.2 Potential African Supply Bases

There are two main centralized supply bases in Eastern Africa. The Nairobi, Kenya, base will be set up for most land-based supplies. The special executive housing and project offices on the island will support the coordination of all the ocean-going freight.

11.3.3 Opportunity Island

Opportunity Island is located approximately 20 miles (32 kilometers) off the east coast of Kenya, Eastern Africa. It is the very upper and somewhat separated northern part of the Tanzania Zanzibar Archipelago. Opportunity Island is noted for its rich rainforests, pristine water, fabulous coral reefs, and excellent sport fishing.

11.3.4 Asian Fabrication Yards

Several fabrication yards in the Middle East, Asia, and the Far East are being considered for fabricating modules, racks, units, and other structures for the LNG plant and associated facilities. These yards are based in Abu Dhabi and Dubai, UAE, South Korea, and China. There are others of interest in Indonesia and Southeast (SE) Asia.

11.3.5 The Houston Engineering Offices

The detailed engineering design for the facilities will be led out of the successful bidder's US corporate offices. It is anticipated that the successful bidder will also use one or more low-cost support offices out of either Mexico, India, or Indonesia.

11.3.6 Project Scope

The downstream scope of work for the project includes the construction of the two LNG trains and LNG storage and load-out facilities, including a jetty, the domestic gas plant, utilities, a material offloading facility (MOF), and a 2000-construction worker camp-type village.

The design of the domestic gas plant facility features improvements in greenhouse gas emissions performance, including improved waste heat recovery and the provision of facilities for the injection of reservoir pressure carbon dioxide deep into the sub-surface.

The design effort is to be a modular construction strategy to minimize the impact of the construction effort on the final site location and surrounding areas. It is anticipated to be conducted from the successful bidder's main operating center located domestically in the US, with additional support from global engineering centers in Mexico, India, and/or Indonesia. The plan is to use one or more fabrication yards across Southeast Asia and the Far East to support the planned 120,000 tons of LNG modules.

11.3.7 Project Characteristics

- The expected total installed cost of the project: $10 billion with the estimated split of work is as follows:
  - Design/engineering: 10%.
  - Module fabrication & transportation: 25%.
  - Major owner long-lead equipment (non-modular, non-EPC): 10%.
  - Equipment and site construction: 55%.
- The target project duration: 4 years.
- Sector: Oil & gas (downstream).
- Location of:

- Owner engineering team: US, potentially Houston.
- Owner site management: Nairobi, Kenya, and Opportunity Island.
- Low-cost engineering center: potentially Mexico, India, and/ or Jakarta.
- Module design office: Houston.
- Module fabricator(s): depends on scope but estimated to two main module fabricators.
- Site: Opportunity Island, offshore East Africa.

Availability of local infrastructure resources is as follows:

- Transportation infrastructure (inland):
  - Truck (road/highway): Limited, two-lane paved or dirt roads.
  - Rail: Limited.
- Transportation infrastructure (coastal):
  - Ship (deep port): Adequate water depth but will require new purpose-built LNG load-out jetty.
  - Barge: New purpose-built construction dock required for bulk materials.
  - Jetty/port for module offload: Adequate water depth but new purpose-built MOF required.
- Power: Self-generated power for both housing and project site.
- Water: Potable water imported or generated via reverse osmosis.
- Sewage: Disposal of all required to be developed.
- Phone lines: Landlines to be installed by successful EPC bidder.
- Telecommunication: Satellite to be installed by Big Oil.
- Housing: Purpose-built by Big Oil (executive compound).
- Temporary housing facilities (aka, Man camp): Built by successful EPC bidder as part of early works prior to EPC.

Other local resources are limited:

- The quantity of the nearby labor market in Kenya is inadequate to completely support the project.

- The quality of the labor market in Kenya is inadequate to support all aspects of the project. Early civil works are an option for local craft utilization.

- The site laydown space: Limited to the mainland or the north part of Opportunity Island. Opportunity Island is too small to support a full stick-built project philosophy.

- The requirements for locally sourced labor: Site grading and civil works (100%).

- The requirements for locally sourced materials: Rock and paving materials available onshore. Island has no construction materials available.

- Main module fabricators and their awarded module scope:

  - Train #1 (60,000 mt) — China.

  - Train #2 (60,000 mt) — Southern China.

  - Pipe racks and utilities — Abu Dhabi, UAE.

- Site labor productivity for Kenya craft relative to Gulf Coast: 5.0.

- Client-supplied or EPC-supplied (pre-issued) equipment or materials:

  - Engineered equipment: All.

  - Steel: All except for module steel, which is provided by the fab yard.

  - Piping: Alloy by EPC/carbon steel by both fab yard and EPC (depending on final destination).

  - Bulks: EPC.

  - Long-lead: Client initiated, novated to EPC at contract award.

- Project difficulties:

  - Weather (extreme): The site is in a cyclone belt.

  - Logistics challenges (transportation of modules): No infrastructure.

  - Environmental impact: Much of the island is either tropical rainforest or surrounded by coral reefs.

  - Labor issues: The site has local labor issues.

  - Regulating impact: Must deal with both the local island government as well as the Kenya national government.

  - Team turnover: Local craft are scarce and tend to have a high turnover of personnel, especially during the tourist season.

11.3.8 Project Stakeholders

- Big Oil Ltd is a leading company in the global oil and gas industry with proven technical and management skills for safe, efficient, and environmentally responsible development.

- The successful EPC contractor will be selected from one of several of the world's premier engineering, procurement, and construction companies. Most with a history of more than a century in the business, these companies have evolved into prominent players in the energy and petrochemicals sectors.

11.3.9 Modularization

- Company's history of use of modularization: All the EPCs have good module experience.

- Quantity and quality of labor where the fabrication facilities are located: All fab yards pre-qualified and selected have an adequate supply of qualified craft.

- Approximate maximum percentage modularization possible on the project (percentage of all site work-hours exported to any/all fabrication shops): 75%.

- Maximum possible direct manual site work-hours moved or expected to move to fabrication shops: Percentage total direct manual work-hours: approx. 75%.

- Indirect, non-manual site work-hours moved or expected to move to fab shops: About 20%.

- Business case drivers for modularization on this project:

  - Schedule

  - Labor cost

  - Labor productivity

  - Labor supply

  - Environmental

  - Site access

  - Site availability

- Differences between labor in the jobsite and labor in the fabrication shop:

  - Labor cost: Assume a local all-in wage rate (AIWR) of $35/hour. Assume imported craft (temporary housing facilities craft labor) AIWR at about $65/hour. Assume Far East (Korean) AIWR at about $65/hour. Assume

Middle East at AIWR of $35/hour. Assume China AIWR at about $45/hour.

- Labor productivity: Local labor productivity factor is 5.0 versus USGC, which is the basis of comparison at 1.0, but in actuality is currently running between 1.2 to 1.5.

- Korea and China have an assumed 1.5 and 1.3 productivity factors, respectively. The Middle East has a 2.0 productivity factor.

- Project design details requirements to facilitate modularization:

  - Fabrication: The project should be specifically designed to be modular.

  - Transport: The project should be specifically designed for ocean-going transportation of the modules.

  - Field installation: The module installation sequence should be incorporated into the path of construction, and therefore the initial design must be completed as noted in the schedule. This requires that early engineering is completed on much of the equipment to be installed in the modules, especially any long lead procurement, like high alloy air coolers that sit on top of the main pipe racks.

- Total number of modules: 35 major modules and approximately 125 smaller PARs and PAUs.

- Sizes of the modules:

  - Largest (by dimension): 30m H 70m L 30m W 4000 metric ton.

  - Smallest: 15m H 24m L 12m W 500 metric ton.

  - Heaviest module: 6000 metric ton.

  - Lightest module: 500 metric ton.

- Types of units/sub-units modularized:

  - Process equipment

  - Utility equipment

  - Loaded pipe racks

- Units duplicated:

  - The project has two trains. As such, numerous modules are duplicated.

11.4 In-Class Exercise

Each of the five project phases is discussed below, providing information that might be typical of the content and level of detail that one would expect to be available at the beginning of each project phase. The student/reader is requested to work through the business case for each of the five project phases. There is more than one potentially correct analysis, so discussions should be encouraged on the pros and cons of each option developed, comparing and analyzing for the best composite—if one can be found.

While there can be more than one "correct" answer, depending on what the student has assumed and how the student has incorporated the information provided into the overall phase planning, there is only one "best" answer. So, in order to make the exercise consistent through the project phases, we, as the authors, have identified "the primary solution" and, based on this, will re-set the exercise to reflect this solution for each of the next project phases. This will be accompanied by explanations of how and why the provided data was analyzed, any other information that may have been identified (or assumed), and potentially some actual data analysis from the phase work. Then, the student is requested to take this "primary solution" scenario and use it as their base case for working through the analysis of the project in the next project phase.

For the purposes of this exercise, we will be following the efforts of one of the EPC contractors: Production Facilities Incorporated (PFI). They are a major EPC contractor that has been around for over a century and are very well versed in modular fabrication. They have a main office in Houston and lower-cost engineering and project support offices in Mexico, Indonesia, and India. They have a solid module savvy organization and a management that is very proactive in terms of supporting the modular option.

At the end of this exercise, the reader will have a complete and consistent scenario of one option on how this project might have played out in real life. As a textbook, the solution is not part of the printed version but available for instructors (not for students) as a separate supplement. Contact Dr. Choi (choi.jinouk@gmail.com) for the solution file.

We now present a summary of the work performed to date by project phase.

OPPORTUNITY FRAMING (FEL-0)

PFI began their study on the modularization potential in the Opportunity Framing (FEL-0) phase. PFI worked with Big Oil to identify the project drivers, cost implications, and revenue potential (with as much information as Big Oil would share on these details).

PFI immediately set forth to work overall options on these alternatives at a very high level to determine if any roadblocks or black swans were lurking in the analysis.

From the FEL-0 analysis, PFI evaluated both the land and island options as well as the option of a floating LNG production and storage system, which would keep the entire facility offshore. In conjunction with Big Oil, PFI ran early alternative options on LNG plant site location (island and mainland)/floating LNG or land-based plant/and further sub-options on several individual plant site locations for each of the island and mainland options.

PFI worked with the upstream planners from Big Oil in terms of options and timing.

PFI then ran some high-level scheduling options. The results provided PFI with the guidelines in terms of the potential module sizing and work-hour movement off the proposed project site (whichever one would be chosen).

11.4.1 Suggested Student Development Activities

The student/reader is requested to work through the business case for this phase framed by the description above:

1. What was PFI's main goal in this phase?

2. What particular items of the business case did PFI concentrate on in this analysis?

   - Perform a very high-level comparison of the estimated cost to build at the site versus what might be expected at a fabrication yard utilizing the economic productivity ratio.

   - Run the CII PPOMF tool and present the result.

3. What are the major phase activities?

4. What are the major modularization goals?

5. What concerns should have been raised in this phase?

6. Were there any preliminary conclusions or results identified or developed?

7. Which execution plan differences are identified in this phase?

8. Which critical success factors should be involved for the first time in this phase?

9. What should PFI do in terms of owner feedback and continuity?

ASSESSMENT (FEL-1)

PFI's work in FEL-0 reached the conclusion that the LNG facility was a viable option and that there was money to be saved by some amount of pre-assembly, along with an idea of the amount of modularization that would be optimal.

By careful examination of the project details, PFI correctly identified both the primary project driver as well as the very close secondary project driver. Because of Big Oil's reluctance to put all of their eggs (and trust) into a single company, they continued to proceed with caution and began a campaign of soliciting information from other engineering sources on the best-suggested path to proceed (in more or less a competition of ideas).

PFI moved into FEL-1, understanding that they would have to compete with another engineering firm Big Oil had worked with in the past. This situation made the collaboration between Big Oil and PFI a bit strained and was somewhat counterproductive. However, PFI continued to move the analysis forward in as collaborative an atmosphere as possible.

In this phase, PFI began evaluating the alternatives, both in terms of the process itself and the overall project layout. They took into consideration the impact of all the costs associated with the various parts of the process and layout options.

PFI began a risk analysis of the project. The aim of this effort was to make sure that all options were at least identified, and risk ranked in more detail.

PFI also looked into the magnitude of the potential modularization options in terms of percentages of total site work-hours removed. As a result of these analyses, PFI determined (and confirmed Big Oil's initial assessment of) the best site location for the LNG plant.

A summary of all these analyses that PFI developed was presented to Big Oil as a recommendation at the end of the assessment phase. There were still some outstanding options to resolve. However, at least by this time, PFI had cobbled together a plausible project execution strategy and juxtaposed the other alternatives in such a manner that it identified the various options along with their faults and benefits and pointed toward the better options.

Big Oil was impressed with PFI's thorough approach and agreed to allow PFI to continue their study efforts into the project selection phase (or pre-FEED). Of course, PFI was ecstatic, but after a short celebration, they began to assimilate the larger group needed to develop the deliverables for pre-FEED.

11.4.2 Suggested Student Development Activities

The student/reader is requested to work through the business case for this phase framed by the description above:

1. What was PFI's main goal in this phase?

2. What particular items of the business case did PFI concentrate on in this analysis?

   • Conduct the business case analysis using the business case analysis tool and present the result.

3. What are the major phase activities?

4. What are the major modularization goals?

5. What concerns should have been raised in this phase?

6. Were there any preliminary conclusions or results identified or developed?

7. Which execution plan differences are identified in this phase?

8. Which critical success factors should be involved for the first time in this phase?

9. What should PFI do in terms of owner feedback and continuity?

SELECTION/PRE-FEED (OR FEL-2)

In this selection phase, PFI needed to review all the alternatives identified in the assessment phase with respect to all the potential project variations in terms of equipment, process, layout, delivery, cost, schedule, fab yards, etc., to reduce them down to the best options for each.

As PFI wrapped up their pre-FEED efforts, discussions began regarding the opportunity for Big Oil to reduce the project schedule by several months. This schedule savings could be realized if Big Oil continued with PFI into the FEED instead of competitively going out for bids on the FEED. So, PFI challenged Big Oil on the subject, arguing that not only was there a significant loss of continuity, but the time to develop the FEED bid package would eat up most if not all of the potential schedule to be saved.

Big Oil's project leads, impressed by PFI's work to date, were inclined to continue with PFI. However, Big Oil's upper management resisted for various reasons, including not wanting to put "all their eggs in one basket," still not trusting the concept that a single contractor can provide the needed benefits, and a reluctance to deviate from a well-established but dated culture. Instead, they insisted on a design competition, where a second company (similar in size and experience to PFI) was brought in to perform a parallel detailed design effort. The aim of this design competition was to obtain an alternative package from the other engineering firm that, in theory, would be substantially different, lower cost, with a faster schedule to completion. The unstated premise for Big Oil's push of the design competition was that as a competition, it would force the two competitors into developing a better product than what either would have developed individually. While PFI disagreed with the competition and its basis, PFI knew that this bit of Big Oil culture would not be changed.

Of course, the downside of this competition was the project schedule loss of 3 to 6 months in terms of the bidding, evaluation, and award, as well as the lack of continuity due to the incomplete "sharing" of PFI's pre-FEED data to their competing engineering company.

Because of PFI's extraordinary efforts to date, Big Oil offered PFI the option to become Big Oil's third party project management team (PMT) to provide project management assistance to Big Oil on this design competition and later on during the next project phases. But, because this would eliminate PFI from participating in the design competition itself and ultimately the EPC portion of the project, PFI declined. As such, PFI decided to let another third party assist with the project management support of Big Oil while they continued with the design competition in hopes of winning it and continuing into the EPC phase (which was where the big bucks could be made . . . or lost).

As was typical, Big Oil offered to pay both engineering companies a set price, a not to exceed amount of money to complete the design competition. This also limited creativity and innovation by both competing engineering teams as the budget forced hard choices on how far to go in maturing some of these value-improving ideas.

PFI felt that they had the upper hand in this competition. They had performed well in the assessment phase, and their team was cohesive and working as a single unit, all with the goal of making the module option a success. They also had subtle cost and schedule saving ideas that would be incorporated into the design competition they had been working on that they were not required to share with their competitor and further decided not to share immediately with Big Oil. This was their competitive advantage, and they wanted to maintain it. Again, by nature, the design competition prevented both teams from working from the most optimal point of project reference.

Of course, like the decision to make this a competitive FEED, the decision of PFI to withhold some of the potential optimization ideas was not in the best interests of the project itself. However, PFI felt it was necessary to maintain a competitive advantage and this was driven by Big Oil's initial reluctance to continue with PFI and their desire to create and support this design competition.

11.4.3 Suggested Student Development Activities

The student/reader is requested to work through the business case for this phase framed by the description above:

1. What was PFI's main goal in this phase?

2. What particular items of the business case did PFI concentrate on in this analysis?

3. What are the major phase activities?

4. What are the major modularization goals?

5. What concerns should have been raised in this phase?

6. Were there any preliminary conclusions or results identified or developed?

7. Which execution plan differences are identified in this phase?

8. Which critical success factors should be involved for the first time in this phase?

9. What should PFI do in terms of owner feedback and continuity?

BASIC DESIGN (FEED) OR (FEL-3)

Knowing they were competing with a favorite third party EPC, PFI initiated discussions with the small Big Oil team that would be overseeing their efforts, ensuring that the Big Oil team would maintain the PFI information confidentiality during the FEED competition.

While assurances were provided that any technical or commercial advantages would not be automatically shared with their FEED competitor, there was still a certain level of distrust and reluctance for PFI to share every advantage they were considering. PFI agreed that while this was obviously not the best condition for developing the absolute best module project option, they would have to maintain a certain level of distance in terms of the real cutting-edge benefits they were planning to incorporate into their version of the FEED package.

As a side note: This was disappointing to PFI as all understood that the path to the absolute best option was to work in a completely collaborative manner between Big Oil and themselves. The current contracting structure made this very difficult, but PFI persevered and did their best to be as collaborative as they felt comfortable being in this situation.

So, with that internal understanding within the PFI organization, PFI began by proceeding in the early part of this phase to complete the development of the optimal option by identifying all remaining (major) outstanding decisions and developing a schedule and sequence for their resolution.

PFI confirmed that the specific optimal module case identified in the pre-FEED study work was the appropriate case to be taken forward to Big Oil to be followed in the detailed design, procurement, and construction of the EPC phase.

PFI identified a possible schedule reducer and decided to follow this option through development with the intention of presenting it as a non-solicited alternative to Big Oil since the schedule was so important.

Also, PFI management took a chance and doubled down on their design efforts toward completion in order to get to a point where the design was essentially complete in terms of which alternatives would be selected. Thus, PFI produced a comprehensive FEED package by, in fact, including a few non-solicited scope items at PFI cost (in order to be better situated if awarded the EPC contract).

PFI presented their proposal to Big Oil, and the waiting game began. Big Oil had a few questions but did not seem overly impressed. PFI, sensing they were about to lose out to their FEED design competition competitor, pressed Big Oil on what Big Oil thought about the unsolicited schedule-saving alternative presented at the end of the proposal. They quickly went over the savings of several months it could provide.

Big Oil only told PFI that they would look into this. Unbeknownst to PFI, there were some internal e-mails back and forth within Big Oil chasing this alternative analysis down. (What PFI found out later was that apparently, this alternative execution plan had been dismissed by one of the Big Oil evaluation employees because it was unsolicited and, as such, did not meet the contracting bid requirements. The employee did not even evaluate the ramifications.)

After a few days of analysis and various reviews of the schedule-reducing option, Big Oil management formally returned to PFI, thanking PFI for the alternative. It turned out that this option indeed provided a better fit for the initial exploitation of the gas field and initial LNG production.

PFI was thanked for their extra efforts in developing the earlier alternative.

Big Oil agreed that PFI was the type of EPC contractor they wanted on their team—a team player that had their best interests at heart and was willing to go beyond what was expected, anticipating issues, and solving them ahead of time and without prompting.

PFI was again ecstatic! They had taken a chance by not accepting the PM role on the field development and, as a result of some heads-up research, some collaboration, and some extra internal efforts in terms of evaluation of Big Oil's unidentified needs, had won the praise of Big Oil and the opportunity to perform the entire EPC effort.

11.4.4 Suggested Student Development Activities

The student/reader is requested to work through the business case for this phase framed by the description above:

1. What was PFI's main goal in this phase?

2. What particular items of the business case did PFI concentrate on in this analysis?

3. What are the major phase activities?

4. What are the major modularization goals?

5. What concerns should have been raised in this phase?

6. Were there any preliminary conclusions or results identified or developed?

7. Which execution plan differences are identified in this phase?

8. Which critical success factors should be involved for the first time in this phase?

9. What should PFI do in terms of owner feedback and continuity?

EPC (EXECUTION, DETAILED DESIGN, PROCUREMENT, AND CONSTRUCTION)

PFI had also taken a chance in the FEED development by including initial efforts on data management and organization. As such, they were now in an excellent position to kick off the EPC portion of the work when it was to be awarded and proceed on productive work much sooner (and with little re-work of previous efforts).

PFI's was awarded the EPC portion of the job, and they began the main activities required to execute a proper EPC, including all the internal and external project group interfacing.

And the rest is history.

11.4.5 Suggested Student Development Activities

The student/reader is requested to work through the business case for this phase framed by the description above:

1. What was PFI's main goal in this phase?

2. What particular items of the business case did PFI concentrate on in this analysis?

3. What are the major phase activities?

4. What are the major modularization goals?

5. What concerns should have been raised in this phase?

6. Were there any preliminary conclusions or results identified or developed?

7. Which execution plan differences are identified in this phase?

8. Which critical success factors should be involved for the first time in this phase?

9. What should PFI do in terms of owner feedback and continuity?

We repeat here: As a textbook, the solution to the exercise is not part of the printed version but available for instructors (not for students) as a separate supplement. Contact Dr. Choi (choi.jinouk@gmail.com) for the solution file.

chapter 12 Standardization

The Holy Grail of Pre-Assembly

If the title doesn't get your attention, we are not sure anything (in this book) will. Standardization is the ultimate goal of everyone involved in modularization or offsite assembly, whether they currently know it or not! It is the next step in the evolution of the planning and development of a project's offsite scope. So, what is standardization?

[It is] the development and use of consistent designs for regularity and repetition.

(Construction Industry Institute, 2019)

In short, "Design one, build many."

Standardization is also the ultimate goal in terms of cost and schedule. Like modularization, the standardization analysis should be considered, regardless of the project scope or what you currently think the current potential is for standardization—just in case there is some potential. So, hopefully, by the end of this chapter, you will understand why the analysis of the standardization potential of your next project should be one of your goals, along with the modularization planning you are currently undertaking (and be as excited as we are about it)! In addition, we will explain who is responsible, when such analysis should be started, what additional efforts one should expect in terms of developing a standardized program, and a couple of things to watch out for.

12.1 Why the Interest?

In addition to all the benefits derived from modularization mentioned to date, standardization provides even more.

12.1.1 Design Benefits

The obvious benefit is the "design one, build many" goal seen with most mass-produced products, from nuts and bolts to automobiles. Anytime you can avoid re-engineering, there is the potential of eliminating that percentage of your efforts on the subsequent projects.

12.1.2 Learning Curve Benefits

Along with eliminating the designing efforts on subsequent projects, all the associated "learning" benefits come with duplicating efforts. For example, for all subsequent projects, everyone associated with these projects has gone through the project development: from procurement to module fabrication, to installation and on-site construction to start-up/operations/maintenance. The result is an optimization of everyone's efforts and typically an increase in efficiency for the second and subsequent projects. Of course, this assumes that these subsequent projects are sequential enough to be able to maintain a continuity of workforce.

12.1.3 Procurement Benefits

Besides the benefits of having purchased equipment and materials before and understanding the associated issues, there is the ability to procure in bulk and in advance, potentially realizing additional cost savings. This also translates to construction materials management and ultimately to the equipment sparing philosophy, where a common set of "spares" can be used to support multiple facilities and their operations.

12.1.4 Cost and Schedule Benefits

All of the above results in an overall program cost and schedule benefits since multiple units can be designed and built faster and be less expensive and operated by an O&M team that understands them better (because they are identical or at least very similar). There are some "extra efforts" required to standardize, but they will make more sense after working through how the standardization opportunity might present itself in the workplace.

12.1.5 Overall Value Captured

O'Connor et al. (2009) demonstrated that four standardized low sulfur gasoline projects saved $56 million (approximately 12.7% of the total installed cost). The Construction Industry Institute (CII) (2019) reported that eight different standardization programs with 45 projects saved 10% of the total installed cost, 25% of the life cycle cost, and 15% of the schedule in their upstream and midstream case projects. These savings were gained due to the benefits of standardization such as "design once, reuse multiple times," "design and procurement in advance," "accelerated response to schedule needs," learning curve benefits in fabrication, construction, commissioning and start-up, and operation and maintenance, "decommissioning cost savings," to name but a few (Construction Industry Institute, 2019).

12.2 How Might It Impact You?

Whether you even think the suite of projects currently being managed by you has even the remotest opportunity to be "standardized," you will eventually be confronted with the situation below, in some form or fashion.

EXAMPLE

The project you are supervising is not a first of a kind but a variation that provides additional benefits in terms of carbon capture and reduction in greenhouse emissions. It is currently under construction in the field, and construction is going well. With all the past issues in the engineering development of this project, you are encouraged by the prospects of actually coming in on schedule and under budget. The boss is happy, the client is happy, and you are ecstatic and, more importantly, ready to move on. Your boss has been keeping upper management apprised of progress, and they also like what they see.

Late Friday, the client provides feedback through business development that they see the additional potential for this new technology and are interested in building three more similar plants. Note that they do not ask for "duplicate" plants since site conditions and meteorological conditions vary. The question from them was: "Can you standardize and how much additional $$ and time can be saved on the next ones?"

Of course, Business Development affirms: "Our company has the capability!" Upper management dictates: "Make it happen!" Your boss says: "Develop the proposal – after all, you have all weekend!"

And with that, your dreams of actually getting off this project anytime near term disappear, and you are tied to the prospect of doing something you had not done before—coming up with a way to build the subsequent plants at a reduced cost and schedule (to maintain the favorable impression your bosses have).

So, how do you approach the planning? What do you assume can be shortened? What can be duplicated? Where do you start? Where is a good guide when you need one?

So, you start your research—maybe a "Google" search—"Standardization Planning Guide."

Good luck there—the search response received was "it looks like there aren't many great matches for this search."

In discussions with fellow workers who have applicable standardization experience, they indicate that it requires even more pre-planning than modularization. So, how do you start from scratch with nothing standardized on the first plant being supposedly duplicated?

So, back to the drawing board (and back to this book and particularly, this chapter).

Fortunately, the problems are not as insurmountable as the above project scenario leads one to believe. But, let's start from the beginning and assume you are working on serial #1 of multiple plants to be standardized, and the decision was to be made upfront. With that as a basis, we can then come back and tackle the late standardization decision developed in the scenario above and how one might handle that.

12.3 How to Implement Modularization with Standardization?

12.3.1 The Two Methods of Combining Modularization and Standardization

The more traditional method is to take a plant and modularize it. Then take the modules and standardize them. What you get are essentially "standard modules" that offer some of the benefits of both procedures, but not all.

The method we suggest is to take that same plant and first standardize it. That is developed such that many of its sub-parts are "standard" in design and manufacture. Then take this standardized plant, including all the small standard pieces and processes in it, and modularize the plant. The result is a standardized (or replicable) plant that has been further optimized to combine the standard pieces that work well together in terms of process into modules. These modules, composed of standard parts, now make up a Modular Standardized Plant or MSP (Figure 12.1).

The additional benefits received are from the efforts involved in first creating standard components that can then be assembled into the most optimal module solution for a project. It is this version of standardization that we recommend and assume will be part of any standardization process.

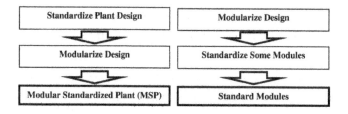

Figure 12.1 The two methods of combining modularization and standardization. Source: O'Connor, O'Brien and Choi (2015). Reproduced with permission of American Society of Civil Engineers.

12.3.2 Ten Factors to Determine Potential for Standardization

So, where does one start?

CII RT-UMM-01 Standardization research (Construction Industry Institute, 2019) provided 10 level-1 factors to be analyzed to determine the potential for standardization. These are listed below in the order provided in the research documents:

1. Management Commitment to Standardization

2. Development Plan

3. Degree or Level of Standardization

4. Market Analysis

5. Standardization Process and Work Process

6. Economic Evaluation

7. Number of Standardized Projects/Units

8. Project Execution Plan

9. Technology Maturity

10. Marketing

12.3.3 Suggested Standardization Study Sequence

Below is a suggested initiation sequence for analyzing these 10 important standardization potential factors. While each project may differ, one cannot go wrong by using this sequencing as an initial basis for such an analysis. Which of the above factors should be evaluated first? The market analysis, of course!

12.3.3.1 #1 The Market Analysis

The market survey identifies the world's appetite for this product. It is the basis on which the entire standardization program is based, with the ultimate goal of satisfying the world's appetite for the product being marketed.

Of course, it is unrealistic to assume that your plant or process, etc. can fulfill all the world's needs. So, the effort becomes the vetting of the standardization potential of a given project in terms of your company's capacities and goals. This effort becomes the process of identifying the optimal potential for standardization with respect to your specific project, company appetite, capacity, long-term goals, etc. The market analysis becomes the standardization equivalent of what the business case 13-step analysis was for the optimal module scope. But the standardization market analysis looks completely different.

The goal of the modular business case analysis was to determine the optimal module configuration for a project. The market analysis's goal is to determine how many similar units one can make and how close they can be to absolute duplicates of the first or prototype. For, as you strive for higher and higher levels of duplication, some trade-offs must be made with respect to sizing, capacity, variability, uniqueness of design, and everything else that makes a plant "special and unique." All the things that each customer typically expected in terms of an exact fit to what is wanted.

If the customer is not going to get exactly what they want, there must be enough benefits, such as cost and schedule savings, so that the customer is OK with accepting something slightly different than what they were asking for.

That is why the market survey is so critical. All this must be very carefully discovered during the market survey, or you run the risk of developing a product that can be duplicated but very few people, if anybody, will purchase and you are stuck with a very expensive and potentially unmarketable prototype.

The following is a list of the key ingredients of a market survey (Construction Industry Institute, 2019; Choi, Shrestha, Shane, et al., 2020):

- How many?

- How much?

- Margin

- Timing

- Geographic location

- Market size

- Market share

- Market potential to grow

- Volatility

- Risk

- Long-term commodity price prediction

- Find out what you need to know about your customer profile.

Note that the survey hinges on quantities/timing/potential customer base/growth/risk. And each of these items hinges on how well the duplicated product can be sold "as is" without changes. In effect, the "one-size-fits-all" product.

The results of the market survey then feed the following other nine factors in the order shown (at least in the opinions of the authors) to be considered in the standardization analysis.

12.3.3.2 #2 The Development Plan

The development plan takes the data from the market survey and, based on the specifics of the parent company, determines how and when the product will be developed and taken to market and how similar each unit will be able to be produced. Many variables are involved in such planning, starting with capital availability, market share captured, impact to current project plans, impacts to existing resources, etc. These get bounced back and forth, hopefully with a resulting agreement on the best way forward.

12.3.3.3 #3 Degree or Level of Standardization

This is the result of taking the market analysis, which provides an idea of how much variation the customer can accept and determining how that resulting product (and any variability) best fit into the current organization in terms of development, production, and sales. The resulting two considerations evaluated concurrently: (1) market share loss with increased percentage duplication versus (2) loss of profitability with decreased duplication (increased customization) will provide an idea of how much duplication versus customization will be required for the proposed program and, by default will define what gets standardized. To put it simply:

1. the more you standardize (duplicate), the fewer customers who will fully accept the product; BUT

2. the more you customize, the lower your potential profit from each unit.

The degree of standardization is the optimal level that meets both considerations.

12.3.3.4 #4 Management Commitment to Standardization

This is your first upper management review. The high-level analysis of the above three factors will provide the framework for a proposal to upper management. And this would probably be the best time to initially approach them. The management wants a good short story with a background on the level of market interest, market share, as well as impacts on current operations and capital funding. They want to understand the standardization "carrot" in terms of its impact on current company operations.

Further study into the development of more detail on the potential standardization program may be wasted if upper management is not interested in work and conclusion to date. However, if they show interest, now is also the time you casually mention some of the commitment details that will come with such a standardization effort, such as organization support, incorporation into company culture, upper management champion, etc. (to confirm that the interest remains). If so, move to the next factors in order of importance.

12.3.3.5 #5 Economic Evaluation

There was an initial assumption made as part of the presentation to upper management on #3 Degree of Level of Standardization. This needs to be bracketed in terms of standardizing a higher percentage of the product as well as a lower percentage based on the company operating specifics. The point of these alternative studies is to determine if there is a breakpoint in terms of company profit, market share, or some other limiting factor that would suggest backing off from that level of duplication or, conversely, moving up to the next level of complexity.

12.3.3.6 #6 Technology Maturity

This analysis should follow in quickstep with the Economic Evaluation since as the percentage of duplication and complexity increases, the potential for changes in the technology also increases, making the product more susceptible to going out of date or becoming obsolete. The analysis should hopefully confirm that the technology

is sound but not so old that alternate technologies are already beginning to emerge that might impact the applicability of the product.

These two efforts (#5 and #6) will provide the analysis to either confirm or adjust the standardization assumptions made and presented to upper management in terms of suitability and sustainability of the proposed product for the market share anticipated.

12.3.3.7 #7 Number of Standardized Units

This has been identified and analyzed in terms of economics, impact to current company operations, capital requirements, etc. So, the next step is to take this scope and develop the plan to actually produce the quantity required.

12.3.3.8 #8 Project Execution Plan

With the number of units, the previous research into the market, your share, impact on company business, etc., the Project Execution Plan can be developed.

12.3.3.9 #9 Standardization Process and Work Process

As part of this development, efforts will need to be made on determining what standardization work processes are currently available within the company as well as from the outside that can assist in identifying and supporting all the changes and up-front efforts involved in developing the standardization program. These resources will be critical in the development of the standardization program.

12.3.3.10 #10 Marketing

Somewhere in the above mix of the last three factors, the company needs to start actually marketing the product they will be manufacturing. All the correct market analyses in the world will do no good if no one knows the standardized product is available from your company by a certain date. And along with this marketing comes the identification of the main customers and how best to make sure that they receive the best service as well as identify the future markets that will help fill in the gaps/create new future capacity/or whatever the current business case for this product entails.

If you are interested in learning more about these standardization decision-making factors, check (Construction Industry Institute, 2019) or (Choi, Shrestha, Shane, *et al.*, 2020).

12.4 The Standardization Work Process

This process is divided into two parts: The design and development of the prototype (or first unit) and then the production of all the subsequent units. They will be discussed in this order and will help you understand the importance placed on the study sequence presented above.

12.4.1 The First Unit (or Prototype)

This unit is developed similarly to any regular custom-build project that will be modularized. All of the business case efforts associated with a modular approach to a project mentioned in the previous chapters of this book as well as the 13-step business case, will apply. But, in addition to the above, what is different with the standardized project is that there needs to be a concerted effort with respect to identifying the market for these potential identical units/ processes/widgets. This is the market analysis mentioned earlier. Of course, in addition to step #1, the market analysis, are the other nine steps needed to turn the results of the market analysis into your company's reality. This all happens in the very early opportunity framing and assessment phases of a typical project. This additional effort makes these two project phases very important for the project since they identify the alternatives, not only the first project or unit but all subsequent projects or units in this standardization program (Figure 12.2).

The next phase—selection—is probably the most important phase in terms of the standardization process as this is where all the alternatives get distilled down into the best option for standardization in terms of the product being produced, timing, quantity, resulting in impacts on your company, potential impacts in the market, and its profitability. This is where the details are worked out in terms of the mechanics of procurement, engineering, modularization, site construction, and O&M. This sets the stage for how all the subsequent units will be produced. It also sets up a plan for going back and re-evaluating the standardization process in terms of any potential future impacts from design, technology, market, etc., that might appear in order to avoid a situation where the product being produced is no longer in demand. This is commonly called "lessons learned."

This additional work in these first phases of the first or prototype product is one of the "extra efforts" mentioned earlier when evaluating the potential for standardization. It will take longer and be more expensive because of the involved market survey and additional analysis of how best to develop and market the product. But the nice thing about this is that it will be the only time you spend this extra effort on the program if done correctly.

12.4.2 All Subsequent Units

If the project tasks for the initial or prototype project are performed properly, the second and subsequent projects/ unit efforts begin with an abbreviated basic design or front end engineering design (FEED) phase (Figure 12.3). And even in this phase, the tasks have been essentially completed in terms of the project design. What typically remains for this basic design phase is the validation of the design criteria in terms of the specific location, feed inputs, soil conditions, etc., and perhaps some minor adjustments to how the unit or project interacts with its environment in terms of support, weather, etc.

Because the project phases required to take project #2 and all subsequent projects or units have been abbreviated by the removal of opportunity framing, assessment, selection, and even parts of basic design, the engineering, planning and procurement phases can be completed much quicker and with less cost per project or unit. Construction

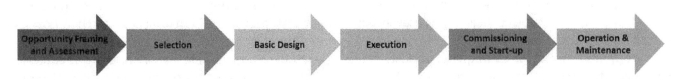

Figure 12.2 Project phases for the first unit.

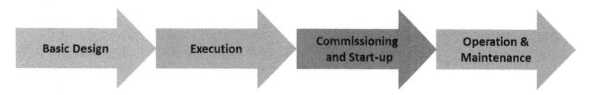

Figure 12.3 Project phases for second and subsequent units.

and start-up also develop some efficiencies and typically are able to start and complete quicker. Finally, O&M has become familiar with the equipment and processes, so the commonality among all the units makes both the operations as well as any maintenance more "routine."

12.4.3 Owner/Client Accountability

One other key difference between a standardized program and other projects is the importance of the owner/client in terms of responsibility. Similar to the modularization case, where the owner plays a large part in the early decisions that make modularization a success, because the standardized approach is based on duplication, in many cases, this requires a super-concerted effort on the part of the owner/client to develop the market case for the units or products. And to ensure success, most of these important decisions must be made in the early project phases of the first prototype (opportunity framing, assessment, and selection).

The EPC contractor has their own set of responsibilities, but these pertain to ensuring that the team's continuity and the approach are maintained throughout the program of production. This brings us back to the initial situational example presented at the beginning of the chapter—what to do when faced with the need to develop a standardized program after the fact.

12.5 The Standardized Approach: Critical Success Factors for Standardization

There are certain actions, similar to the critical success factors that we identified for implementing a modularization solution approach, that are critical to the standardization solution. These are not all-inclusive but are considered by the authors as important enough not to be ignored in this high-level summary discussion. Since the owner/client has already decided in terms of what is to

be standardized, it falls on the EPC contractor to identify how best to maximize cost and scheduling (to provide the best solution for the client and be competitive in the marketplace).

12.5.1 Continuity

The first is continuity within the program being developed. As mentioned, standardization offers the benefits of developing learning curves for multiple groups and activities within each project of the program. But this works best when the "teams" can be maintained throughout the program. So, very early, the EPC contractor needs to identify how they will maintain this team essence or core that will be able to use and incorporate the learning throughout the program.

12.5.2 Alignment

Following the continuity is an overall general program/project alignment from upper management down to the rank-and-file workers at the site. Everyone needs to understand the greater goal in terms of the program and their part in how to make it successful.

12.5.3 Discipline

This is a "standardized" program. That means it must remain duplicatable. Changes to the basic design and implementation are to be avoided "at all costs," as the saying goes. This needs to be a strong attribute within all the teams to maintain this aversion to "change."

So, for the average project manager who has been asked to develop a plan to build three more very similar plants, they must analyze their company's ability to provide a team that will follow these attributes. A review of this simplified approach may also trigger potential feedback to the owner/client in terms of optimization of the program. For example, if the owner/client can optimally time these subsequent developments, there may be incremental cost and schedule savings based on the planning and execution and learning curves developed.

12.6 Conclusion: You're on Your Own

Going further into the various potential scenarios for this proposed example development is beyond this very elementary introduction to standardization. But it is hoped that this chapter has at least identified the potential benefits of standardization, who is responsible, when such analysis should be started, what additional efforts one should expect in terms of developing a standardized program, and a couple of things to watch out for.

For additional details on the facility standardization, we suggest that the reader check out the CII research: RT-UMM-01 Facility Standardization for Program Success (Construction Industry Institute, 2019, 2020) or Dr. Choi's academic papers on facility standardization (Choi *et al.*, 2018, 2022; Construction Industry Institute, 2019, 2020; Choi, Shrestha, Kwak, *et al.*, 2020; Choi, Shrestha, Shane, *et al.*, 2020; Shrestha *et al.*, 2020a, 2020b; Choi *et al.*, 2021a; Choi *et al.*, 2021b; Shrestha *et al.*, 2021). The articles include standardization topics on barriers and challenges (Choi *et al.*, 2018), decision-making model (Choi, Shrestha, Shane, *et al.*, 2020), critical success factors (Choi, Shrestha, Kwak, *et al.*, 2020; Shrestha *et al.*, 2020a), timing of critical success factors' implementation (Shrestha *et al.*, 2020b), recipes for standardized projects' performance success (Shrestha *et al.*, 2021), benefits and trade-offs (Choi *et al.*, 2021b), work process and optimization (Choi *et al.*, 2021a), and industry maximization enablers (Choi *et al.*, 2022).

Acknowledgments

The authors would like to acknowledge that some information, ideas, and concepts in this chapter are adopted or duplicated from the reports (Construction Industry Institute, 2019, 2020) conducted by the Research Team (RT-UMM-01) Standardization, as both authors were members of the team. However, the readers may note that the authors introduced only the key or high-level information in this book and interpreted the findings from (Construction Industry Institute, 2019, 2020) based on their expertise and views. Thus, there is some discrepancy between what we presented here versus (Construction Industry Institute, 2019, 2020). We highly recommend that readers check the original CII reports (Construction Industry Institute, 2019, 2020) for details.

References

Choi, J.O. *et al.* (2018) Achieving Higher Levels of Facility Design Standardization in the Upstream, Midstream, and Mining Commodity Sector: Barriers and Challenges. In *Construction Research Congress (CRC) 2018* pp. 278–287. Reston, VA: American Society of Civil Engineers (ASCE). doi: 10.1061/9780784481301.028.

Choi, J.O. *et al.* (2021a) Facility Design Standardization Work Process and Optimization in Capital Projects. In *Proceedings of the Canadian Society of Civil Engineering Annual Conference 2021* pp. 491–503. Surrey, British Columbia, Canada: Canadian Society of Civil Engineering.

Choi, J.O. *et al.* (2021b) Exploring the Benefits and Trade-Offs of Design Standardization in Capital Projects. *Engineering, Construction and Architectural Management*, 29 (3), pp. 1169–1193. doi: 10.1108/ECAM-08-2020-0661.

Choi, J.O. *et al.* (2022) Facility Design Standardization: Six Solution Pieces and Industry Maximization Enablers. In Jazizadeh, F., Shealy, T., and Garvin, M.J. (Eds.) *Construction Research Congress (CRC) 2022*, pp. 491–503. Reston, VA: American Society of Civil Engineers (ASCE). doi: 10.1061/9780784483978.073.

Choi, J.O., Shrestha, B.K., Kwak, Y.H., *et al.* (2020) Critical Success Factors and Enablers for Facility Design Standardization of Capital Projects. *Journal of Management in Engineering*, 36(5), 04020048. doi: 10.1061/(ASCE) ME.1943-5479.0000788.

Choi, J.O., Shrestha, B.K., Shane, J.S., *et al.* (2020) Facility Design Standardization Decision-Making Model for Industrial Facilities and Capital Projects. *Journal of Management in Engineering*, 36(6), 04020077. doi: 10.1061/(asce)me.1943-5479.0000842.

Construction Industry Institute (2019) *Achieving Higher Levels of Facility Standardization in the Upstream, Midstream, and Mining (UMM) Commodity Market-Volume 1: Four Solution Pieces*. Austin, TX: The University of Texas at Austin: Construction Industry Institute. Available at: https://www.construction-institute.org/resources/knowledgebase/knowledge-areas/project-program-management/topics/rt-umm-01.

Construction Industry Institute (2020a) *Achieving Higher Levels of Facility Standardization in the Upstream, Midstream, and Mining (UMM) Commodity Market-Volume 2: Business Case Analysis Model and Work Process*. Austin, TX: The University of Texas at Austin: Construction Industry Institute. Available at: https://www.construction-institute.org/resources/knowledgebase/knowledge-areas/project-program-management/topics/rt-umm-01.

O'Connor, J.T. *et al.* (2009) Executing a Standard Plant Design Using the 4X Model, *Hydrocarbon Processing*, 88(12), pp. 47–53.

O'Connor, J.T., O'Brien, W.J., and Choi, J.O. (2015) Standardization Strategy for Modular Industrial Plants. *Journal of Construction Engineering and Management*, 141(9), 4015026. doi: 10.1061/(ASCE) CO.1943-7862.0001001.

Shrestha, B.K. *et al.* (2020a) How Design Standardization CSFs Can Impact Project Performance of Capital Projects. *Journal of Management in Engineering*, 36(4), 06020003. doi: 10.1061/(ASCE)ME.1943-5479.0000792.

Shrestha, B.K. *et al.* (2020b) Timings of Accomplishments for Facility Design Standardization Critical Success Factors in Capital Projects. In *Construction Research Congress 2020: Project Management and Controls, Materials, and Contracts – Selected Papers from the Construction Research Congress 2020*, pp. 898–906. doi: 10.1061/9780784482889.095.

Shrestha, B.K. *et al.* (2021) Recipes for Standardized Capital Projects' Performance Success. *Journal of Management in Engineering*, 37(4), 04021029. doi: 10.1061/(ASCE) ME.1943-5479.0000926.

Innovative Technologies for Modularization

13.1 Introduction

The use of modern and innovative technologies in modular projects disrupts the conventional way of doing things by facilitating the reduction and eventual elimination of older techniques in preference to newer and more efficient ones, that can improve cost performance and allow for extended and improved productivity. When modularization is combined with innovative technologies, the potential result is a leveraging opportunity surpassing the benefits taken individually. However, there is a historic reluctance in the construction industry to adopt new and innovative technologies, and modular projects are no exception.

This chapter introduces some of the innovative technologies applicable to a modularization application, explains in which phases they can be implemented, and who the primary beneficiaries are. These technologies can be grouped into three areas: (1) visualization, information modeling, and simulation; (2) sensing and data analytics for construction; and (3) robotics and automation. The chapter concludes with a vision of how these could be combined and used.

13.2 Current List of Innovative Technologies for Modularization

There is great potential for implementing innovative technologies in the construction industry and an equally great need for their implementation. The construction industry has historically exhibited slow progress in terms of embracing and implementing such new and efficient technologies. Some of the challenges hindering such implementation have previously been discussed, but because the construction industry is falling behind in terms of the actual application of these technologies, they have been unable to reap the significant benefits that accompany these applications. Counterpart industries, such as shipbuilding and manufacturing, have accepted the challenge, used modern technologies, and gained substantial benefits. So, why is the construction industry, particularly modular projects (where there is considerable potential to implement technologies), unable to do the same?

In an attempt to address this question, a recent study (Choi *et al.*, 2020) investigated capital projects in the construction industry to identify potential innovative technologies that can help in achieving higher levels of modularization. This research study identified 26 innovative technologies most impactful to modularization and most likely to occur in the next five years. These were identified via a detailed literature review as well as detailed discussions and surveys among experienced subject matter experts on modularization and design standardization in the oil and gas sector.

The following lists the 26 identified potential technologies identified:

1. Facility design standardization

2. Industrial robots

3. 3D printing

4. Autonomous vehicles/autonomous construction

5. Light detection and ranging (LiDAR)/reality capture

6. Drones

7. Building information modeling (BIM) design models

8. 5D BIM

9. Virtual reality (VR)

10. Augmented reality (AR)

11. Wearables

12. Smart glasses

13. Intelligent and automated data collection technology

14. Mobile user interface devices

15. Wireless networks for construction sites (WLAN)

16. Construction simulation technologies

17. Simulation-based virtual commissioning

18. Simulation-based operator training

19. Interim product database (IPD)

20. Automated design

21. Digitized commissioning and handover

22. Digital performance management

23. Capital portfolio management

24. Completion management system

25. Real-time field reporting

26. Materials logistics management.

While some of these technologies are very mature for modularization, others have some way to go. For example, technologies such as building information modeling (BIM) and radio frequency identification (RFID) devices are both more mature and more common. As such, they have been widely used in the construction industry for some time. However, in spite of this, modularization experts believe they are not used as often as they should and should gain wider acceptance and utilization in the coming five years (Table 13.1). On the other hand, technologies such as virtual reality (VR), augmented reality (AR), artificial intelligence (AI), and laser scanning are still maturing and, as such, are not being implemented at high levels. As noted in Table 13.1, VR, in particular, is expected to come into greater use in the future as more explore the many uses and benefits of these technologies on construction projects. The survey result can be found in Table 13.1, which presents innovative technologies and their impact scores (out of a maximum score of 5)

Table 13.1 Innovative technologies and their impact on modularization.

ID #	Innovative technologies	Modularization impact score	Likelihood of occurrence in 5 years
#14	Mobile user interface devices	3.63	4.36
#15	Wireless networks for construction sites (WLAN)	3.51	4.22
#5	Light detection and ranging (LiDAR)/Reality Capture	3.37	4.14
#1	Standardization	4.19	4.1
#6	Drones	2.64	4.02
#16	Construction simulation technologies	3.7	3.88
#13	Intelligent and automated data collection technology	3.76	3.86
#26	Materials logistics management	3.93	3.85
#7	BIM design models	3.6	3.78
#18	Simulation-based operator training	3.07	3.6
#25	Real-time field reporting	3.3	3.6
#21	Digitized commissioning and handover	3.08	3.41
#22	Digital Performance Management	3.05	3.31
#20	Automated design	3.83	3.17
#9	Virtual reality (VR)	3.05	3.14
#17	Simulation-based virtual commissioning	3.28	3.12
#23	Capital portfolio management	2.69	3.08
#11	Wearables	2.41	3.02
#10	Augmented reality (AR)	2.99	2.95
#8	5D BIM	3.25	2.94
#19	Interim product database (IPD)	3.49	2.89
#24	Completion Management System	2.52	2.88
#2	Industrial Robots	3	2.74
#12	Smart Glasses	2.38	2.63
#4	Autonomous Vehicles/Automated Construction	2.55	2.51
#3	3D Printing	2.5	2.4

Source: Adopted from Choi *et al.* (2020). Reproduced with permission of American Society of Civil Engineers (ASCE).

for modularization. Table 13.1 is sorted in terms of the likelihood of occurrence of these technologies in the next five years.

In addition, autonomous/automated construction technologies, LiDAR, robotics, and intelligent automated data collection technologies are predicted to impact modularization significantly (Choi et al., 2020).

The use and effective implementation of such digital technologies and their incorporation can very much aid construction projects, especially modular projects, to enhance and improve productivity, cost, and schedule by helping to reduce errors in offsite construction, enhancing safety for onsite and offsite processes, and providing real-time progress status, and evaluating the quality of production in modular projects.

13.3 Technologies of Interest

The impact of the use of technologies to complement modularization is on the rise. More and more, different technologies are being considered for implementation in modular projects, including building information modeling (BIM), 3D modeling, drones, digital twins, mobile user interface devices, construction simulation technologies, automation, and wireless networks for construction sites. In reviewing all the available technologies, we have selected the technologies that have a high impact on modularization but at the same time also have higher potential (Table 13.1) of being used in the future based on the research (Choi et al., 2020). For example, BIM, LiDAR, automation (robotics), virtual reality, construction simulation technologies all have a high impact potential for practical implementation in modularization in the near term. Other literature on technology also corroborates the importance of these selected technologies in the sector of construction (Chui and Mischke, 2019; Construction Users Roundtable, 2020; Ghimire et al., 2021). For further discussion of these technologies, they have been grouped under three headings:

1. Visualization, information modeling, and simulation

2. Sensing and data analytics for construction

3. Robotics and automation.

13.3.1 Visualization, Information Modeling, and Simulation

The use of visualization, information modeling, and simulation technologies such as digital twin in conjunction with AR/VR/MR/XR allows the owners and the stakeholders to better grasp the project benefits and tangibly estimate costs and schedules, promoting the implementation of modularization. Since these technologies enable the users to monitor the real-time progress of projects and simulate hypothetical and potential problems that the project might face, they encourage the owners and stakeholders to review the project in its entirety with tangible results in the early conception and project assessment phases.

Incorporation of digital twin and AR/VR/MR/XR in earlier phases have been observed to be especially useful since one of the major challenges for the implementation of modular construction is obtaining buy-in from the owners and stakeholders in terms of modularization and its benefits on a project. This technology helps get all modular project participants (owners, different contractors, vendors, and engineers) on the same page in terms of understanding the goals and demands of the various project phases, from conception and assessment, through design, construction, and commissioning. Its application should help the users (all stakeholders) better understand and define the scope of the modular projects earlier, in terms of the project phases.

13.3.1.1 Digital Twin

A digital twin is a virtual representation, a model, of the actual project, which allows for the collection of real-world data and information (Goodman, 2019). The digital twin is developed through the use and incorporation of different technologies, for example, sensors. The use of drones, BIM, virtual 3D modeling and simulation, cloud computing, AR, and VR is fast becoming popular in collecting information for digital twins. However, integrating different technologies to actually support the digital twin technology in construction has not matured in modularization and is not being implemented to the levels it should be. When implemented properly, such collected and sourced information allow the users, particularly owners and stakeholders and project team members, to become more involved, addressing potential construction issues before the construction even starts in a project.

Note that the digital twin is not software or a single entity that is to be used in projects. It is an amalgamation of different technologies, as mentioned previously. A digital twin can adopt information from multiple sources, improving by implementing machine learning, AI, and advanced analytics (Goodman, 2019). The result is that users can

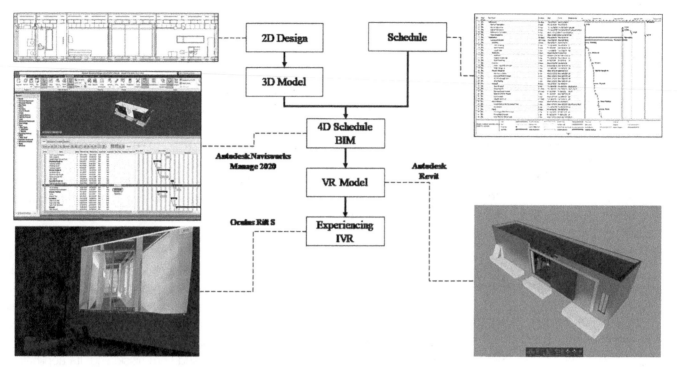

Figure 13.1 Example of 4D BIM schedule model with IVR development process.

better understand the real-time project progress. Further, it empowers users to run simulations of potential and hypothetical situations that may occur in the project design and execution and help address such potential issues and failures.

13.3.1.2 Virtual Reality (VR), Augmented Reality (AR), Mixed Reality (MR), and Extended Reality (XR)

VR is entirely digital or virtual, creating a virtual workplace that allows the users to be immersed completely in real-life workplace scenarios. VR creates a simulated digital world that the user can access using devices (e.g., smart glasses and VR gears). Augmented reality (AR) is a technology used in the construction industry that generates designs and incorporates them into the actual work sites and physical structures at the site. Mixed reality (MR) is a combination of virtual and real-life elements. It is a hybrid where the virtual and the actual can coincide and interact. Extended reality (XR) is a combination of VR, AR, and MR that incorporates all the characteristics of these technologies in one device. XR is still in the early phases of development and will require time to mature in its application in the construction industry.

AR/VR/MR/XR technology allows the users to interact and investigate with the construction sites and identify (potential) issues, keep track and visualize schedules, as well as enhance safety at construction sites. Its use, coupled with creating digital twins for existing projects or future projects, would exceptionally provide a highly immersive experience for the users by merging the real and virtual worlds. AR/VR/MR/XR technology has not matured to its fullest extent in the construction industry, especially in modular projects. In addition, these technologies, especially VR and AR, are often deployed too late in the EPC phase, when more value can be earned by deployment in earlier project phases.

The example use of VR technology in the rendering of a virtual model can be seen in Figure 13.1, which shows the generation process of the 4D schedule using the 3D model of a modular house project. As seen in Figure 13.1, a 4D BIM schedule can be generated by combining a baseline schedule with a 3D model. In addition, immersive virtual reality (IVR) can be used in modular projects to see a virtual representation of modules at different fabrication/construction stages. A 4D BIM schedule can help modular projects reduce risks, particularly in the initial planning and design phases in the modular projects. The detailed development process and the benefits of a 4D BIM schedule model with IVR are explained in (Ghimire et al., 2021).

13.3.2 Sensing and Data Analytics for Construction

Sensing and data analytics technologies enable participants of modular projects to do the following:

1. collect data (in real-time);

2. analyze the collected data;

3. understand the status of the project, modules, or materials;

4. document as-built;

5. help the users to make decisions.

These technologies should be used in combination during the entire lifecycle of modular projects. They will help the owners understand the benefits of modular construction in their projects and encourage them to better understand the long-term gains in cost, schedule, quality, and safety in the projects. Using these technologies in the early phases of modular projects helps stakeholders gain more knowledge about the products from previously executed modular projects and designs, therefore helping them make optimal design decisions. Moreover, using these technologies together can provide for advanced data processing, analytics, and machine learning. The result is better project control along with instant and real-time use of field data and better clarity in project execution, quality control, and schedule analysis.

13.3.2.1 Point Cloud Modeling

Geometric quality inspection of modules is critical. Poor geometric quality (alignment) can cause re-work and disrupt the schedule, especially for multiple prefabricated units that must be connected at the site. Such poor alignment can affect productivity in installation and other connection efforts at the site. Historically, a dimensional inspection of prefabricated units is performed by hand, for example, with measuring tapes. Accuracy becomes especially difficult and cumbersome if the dimensions of the prefabricated units are large, significantly off grade, and if there are multiple units to inspect. An advanced and technologically superior method is required instead of such manual inspections.

For modular construction, a coordinate system, also called 3D point cloud data modeling, can be used to define geometrical or spatial information, for instance, a prefabricated module unit. Point cloud-generated models not only can be used for quality inspection and as a transportation aid but also can be used for planning, feasibility, predictive tests for failures and potential issues, and risk assessments. Data collection for point cloud modeling can be done using LiDAR, laser scanning, photogrammetry, or videogrammetry. Laser scanning or LiDAR is the most accurate point cloud data collection method as it is automated and has high remarkable accuracy and resolution. Collecting 3D data when assembling modular units can help identify issues with the modules during fabrication and assemblies of the modules (Nahangi and Haas, 2016). Accurate 3D points are collected via photos and laser scanning of the modules and generate point cloud models (Nahangi and Haas, 2016). An example of such data collection from an actual "structural frame and pipe spools" module and the subsequent point cloud model generated from 3D data collected from laser scanning is shown in Figures 13.2 (a) and (b).

Figure 13.2 (a) An actual module of a structural frame and pipe spools; (b) the point cloud model generated. Source: Nahangi and Haas (2016). Reproduced with permission of Elsevier.

Figure 13.3 Point cloud models generated from 3D data collected from LiDAR of modular house projects.

Another example of developing point cloud models from 3D data collected from LiDAR on modular house projects is presented in Figure 13.3.

Table 13.2 summarizes the comparison among five different data acquisition approaches to the 3D point cloud in terms of equipment costs, measurement accuracy, maximum measurement range, and example of equipment. The users may select the appropriate approach depending on the budget, purpose of use, and size and complexity of modules. Obviously, the more accurate, the higher the cost.

13.3.2.2 Monitoring Sensors

A combination of different sensors can be used to aid module lifting and transportation in projects. For example, such a monitoring system could include cameras and sensors (such as 360° high-resolution VR cameras), dash cameras, impact recorders, and data loggers. These sensors will collect numerical and visual data. By combining sensor use and locations, impact and damage during module lifting and transportation can be monitored. Figure 13.4 shows an example of visual and numerical data collected by 360° high-resolution VR cameras, dash cameras, and impact recorders during module transportation and lifting.

13.3.3 Robotics and Automation

Robots are predominantly used in the manufacturing sector, where the robot can be relatively stationary and the work brought to the robot. The use of robotics in such production has been proven to reduce costs, save time, improve quality, and reduce risks. Similar benefits could be seen in the modular construction industry. Unfortunately, robotics is not implemented widely in the construction industry, mainly due to the challenges of getting the work to the robot and the general nature of the construction industry's one-off uniqueness.

However, in modular construction, robotics can be used to produce, assemble, and install modules on a large

Table 13.2 Comparison between different point cloud data acquisition approaches.

Approach	Equipment cose	Measurement accuracy	Maximum measurement range	Examples of equipment
3D laser scanning (TLS)	50,000–100,000 USD	1–10 mm [57–60]	70–1000 m [57–60]	FARO TLS [18, 20, 28, 30, 34, 61–65], Leica ScanStation [11, 12, 32, 66–71], Trimble GX [31, 72, 73], RIEGL TLS [74–77]
Photogrammetry	200–2000 USD	Relatively 1–10% (usually 10–100 mm) [78–84]	No specific maximum range, but usually within 50 m to ensure reasonable accuracy	Digital camera [78–80, Single-Lens Reflex camera such as Canon EOS 10D [81, 82], Nikon D-80 [83], Nikon D-800E [84], Rollei 6006 [85]
Videogrammetry	200–2000 USD	5–60 mm [49, 86]	No specific maximum range, but usually within 50 m to ensure reasonable accuracy	Stereo camera set with two Microsoft LifeCam NX-6000 web cameras [49], Canon Vixia HF S100 [86]
RGB-D camera	~150 USD [87]	Relatively 1–6%, absolutely 7–36 mm [51]	4 m [88]	Microsoft Kinect [50–53]
Stereo camera	~3500 USD [89]	1.25% of range [54]	No specific maximum range, but usually within 50 m to ensure reasonable accuracy	Bumblebee® XB3 stereo vision system [54]

Source: Wang, Tan and Mei (2020). Reproduced with permission of Springer.

Gathered Visual Data Example

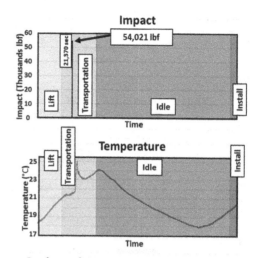

Gathered Numerical Data Example

Figure 13.4 Example of visual and numerical data collected via sensors during module transportation.

scale as more standardized designs and standardized modules are developed. In particular, robotics and automation technologies can

1. improve quality by reducing human errors;

2. substantially increase the speed of fabrication and assembly of modules.

Automation needs to be incorporated into the design, data exchange, and interoperability between different phases in the lifecycle of the project, supply chain, and materials management to garner cost benefits, continuous and efficient exchange of information, and a streamlined supply of materials and finished products. For prefabricated modules, previously robotic arms have been used to fabricate concrete floors (Bock, 2004), produce prefabricated glass reinforced cement panels (Peñin *et al.*, 1998; Pastor *et al.*, 2001), and other items successfully. See Figure 13.5 for an example of robotic arms used for modular construction.

13.3.3.1 Cyber Computational-Physical Systems

One intriguing idea that can be implemented to support modular construction at the job site is the cyber computational-physical systems introduced by Yang, Pan, and Pan (2019). In an attempt to reduce labor at the job site, these authors proposed using cyber computational-physical systems that implement automated robotic cranes, automatic guided vehicles, wireless charging

Figure 13.5 Automation and robotics being used for modular construction. Source: Vasey et al. (2020) Reproduced with permission of Springer Nature / CC BY 4.0.

systems, embedded sensor networks, patrol robots, diverse monitoring sensors, BIM, and cloud systems. This system can support lifting, transporting, and installing modules. Yang, Pan, and Pan (2019) claimed that these systems can improve safety, quality, schedule, and productivity performance. An example of such a system is illustrated in Figure 13.6.

As noted, there is significant potential for improving modular projects via the utilization of modern and innovative technologies currently on the market. The challenge is in the creative way in which each of these technologies is identified, adjusted, and incorporated into the current conventional way of doing things. It is up to the current and next-generation industrial construction workforce to

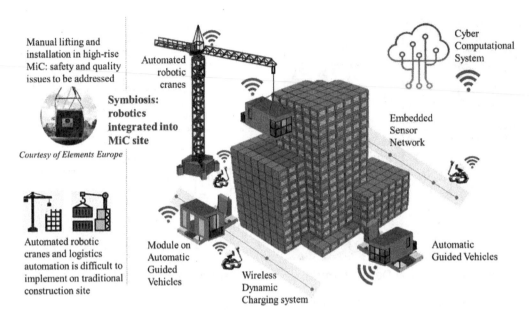

Figure 13.6 Cyber computational-physical system for modular construction. Source: Yang et al. (2019) / Reproduced with permission of Elsevier.

accept the challenge and, in doing so, effectively leverage this technology for the improvement of our industry.

References

Bock, T. (2004) Construction Robotics and Automation: Past-Present-Future. In *World Automation Congress*, pp. 287–294. Available at: https://ieeexplore.ieee.org/stamp/stamp.jsp?tp=&arnumber=1438566&tag=1 (Accessed: 20 March 2022).

Choi, J.O. *et al*. (2020) Innovative Technologies and Management Approaches for Facility Design Standardization and Modularization of Capital Projects. *Journal of Management in Engineering*, 36(5), 04020042. doi: 10.1061/(ASCE)ME.1943-5479.0000805.

Chui, M. and Mischke, J. (2019) The Impact and Opportunities of Automation in Construction. | McKinsey. *Voices on Infrastructure* – McKinsey & Company. Available at: https://www.mckinsey.com/business-functions/operations/our-insights/the-impact-and-opportunities-of-automation-in-construction (Accessed: 20 March 2022).

Construction Users Roundtable (CURT) (2020) *Top Construction Trends for 2020*. Available at: https://www.curt.org/2020/01/07/top-construction-trends-for-2020/ (Accessed: 20 March 2022).

Ghimire, R. *et al*. (2021) Combined Application of 4D BIM Schedule and an Immersive Virtual Reality on a Modular Project: UNLV Solar Decathlon Case. *International Journal of Industrialized Construction*, 2(1), pp. 1–14. doi: 10.29173/IJIC236.

Goodman, J. (2019) Tech 101: Digital Twins. *Construction Dive*. Available at: https://www.constructiondive.com/news/tech-101-digital-twins/561117/ (Accessed: 20 March 2022).

Nahangi, M. and Haas, C.T. (2016) Skeleton-Based Discrepancy Feedback for Automated Realignment of Industrial Assemblies. *Automation in Construction*, 61, pp. 147–161. doi: 10.1016/J.AUTCON.2015.10.014.

Pastor, J.M. *et al*. (2001) Computer-Aided Architectural Design Oriented to Robotized Facade Panels Manufacturing. *Computer-Aided Civil and Infrastructure Engineering*, 16(3), pp. 216–227. doi: 10.1111/0885-9507.00227.

Peñin, L.F. *et al*. (1998) Robotized Spraying of Prefabricated Panels. *IEEE Robotics and Automation Magazine*, 5(3), pp. 18–28. doi: 10.1109/100.728220.

Snide, T.A., Harriman, M., and Schneider, E. (2019) Digital Twin Concept as Applied to Industrial Automation. In *industrial ethernet book*. Available at: https://iebmedia.com/technology/industrial-ethernet/digital-twin-concept-as-applied-to-industrial-automation/ (Accessed: 20 March 2022).

Vasey, L. *et al*. (2020) Physically Distributed Multi-Robot Coordination and Collaboration in Construction. *Construction Robotics*, 4(1), pp. 3–18. doi: 10.1007/S41693-020-00031-Y.

Wang, Q., Tan, Y., and Mei, Z. (2020) Computational Methods of Acquisition and Processing of 3D Point Cloud Data for Construction Applications. *Archives of Computational Methods in Engineering*, 27(2), pp. 479–499. doi: 10.1007/S11831-019-09320-4/TABLES/6.

Yang, Y., Pan, M., and Pan, W. (2019) "Co-Evolution Through Interaction" of Innovative Building Technologies: The Case of Modular Integrated Construction and Robotics. *Automation in Construction*, 107, 102932. doi: 10.1016/J.AUTCON.2019.102932.

chapter 14   Moving Forward

14.1 What's Next?

Hopefully, you got to the end of this publication and immediately started looking for the sequel. But, in case that was not your first thoughts, this chapter is dedicated to a semi-formal wrap-up of the philosophy driving the book along with some "thought-provoking thoughts" on where we as an industry might spend our time and energy in pursuit of a better tomorrow.

With respect to the specific topics covered on industrial modularization, the path and sequencing of the book made perfect sense to us . . . for some of you, perhaps not so much. We tried to walk the thin line of covering basic concepts as well as providing a logical path forward supported by results from research-oriented studies, and all the while "waxing eloquent" on industry-specific initiatives that we found potentially promising.

14.2 So, What Did We Get Out of This?

Besides the not-so-important fact that we (the two authors) proved we could develop such a body of knowledge, we both also actually grew in our understanding of our subject. After all, it is hard to write this much about a subject we have lived with over a large portion of our working careers and not re-examine personal experiences in the light of both the projects themselves and the historical context in which these projects were implemented. The obvious result of this project-by-project re-hash of information was a lot of "20/20 hindsight" and "armchair quarterbacking" (second-guessing)

on how they might have been executed "better." The result (we hope) provided some effective and useful project guidance developed by combining industry research results with personal experiences derived from "boots on the ground." Our aim was that this combination would provide at least one potential path of least resistance for how a project team might navigate through the numerous alternatives presented by the modular option as the various project phases developed.

14.3 What Did You Get Out of This?

Besides a long-winded, somewhat twisted approach to some of the topics and potentially a stiff neck and eye strain if you tried reading too much at one sitting, we hope our discussions and examples provided a logical journey of sorts – from basic module definitions, to early analysis tools and concepts, to actual steps and examples, of how to approach and execute a module project.

We expect that not all chapters were of equal value to you. We also expect you did not agree with some of the ideas and concepts stated. But we do hope that you were able to see past our methods on the approach and over-simplified examples to the reasons and conclusions behind them.

Much of the book can be summed up as an exhaustive explanation of what "good and proper project planning" should be. As such, we hope it has provided some guidance on the proper planning of a modular study or modular project.

14.4 Suggested Future Actions: Our Seven Industrial Modularization Accelerators

We would be remiss if we did not provide our vision of the future. And, like the lessons learned by Scrooge in Charles Dickens' *A Christmas Carol*, our industry is tied to the past, but its future depends on the choices we make now. We cannot change the past, but we can learn from it, save the better parts for re-use, and make sure that we do not repeat the rest.

So here are our **Seven Industrial Modularization Accelerators**, which, if implemented, are what we think will make the future a more module-friendly world:

1. Applied knowledge

2. A different academic teaching approach

3. Identify, acknowledge, and incorporate required paradigm shifts

4. Friendly contracting

5. Industry re-branding (or making the industry sexy, again)

6. More alliances and research

7. Planning techniques and their combinations.

14.4.1 Applied Knowledge

Considered by us to be the most important "success factor." Out of all the critical success factors mentioned throughout these discussions (e.g., module envelope limitation, alignment on drivers, timely design freeze, module fabricator and/or contractor's capability, etc.), **understanding (all) the differences in planning and execution for modular construction** compared to the traditional stick-built approach is the biggest. It sounds simple and obvious, but as always, the devil is in the details. As referenced by the first line of this book's introduction, there are so many aspects of a modular project that need to be planned and executed differently that without a thorough understanding of them all, failure is more likely than not. This is evidenced by the many poorly executed modular projects that lie strewn about the historical industrial construction landscape.

So, how does one develop and implement such an "understanding?"

Read our book, commit it to memory, and execute as described. OK, a bit pompous on our part, but consider that one only gets really good at something that one puts a lot of effort into in terms of research, study, and practical application. The same is true of executing the modular philosophy. One gets out of it what one puts into it. Further, to take this philosophy and actually incorporate it into a modular project, a group (team) of like-minded individuals is required. And for this group to function as a cohesive team, they all must be truly interested in pursuing and reaching the same common goal—in this case, alternative concepts to stick-building.

This will require a little extra effort from all team members that may not be achieved by simply working "9 to 5." Each should be interested in achieving the goal to the point that it becomes a driving force that compels them to inquire/study/practice/play/critique/improve and repeat. But that alone is not enough. It becomes very important that this team be provided the tools needed in terms of background knowledge, not only of the modular philosophy itself, but also of all the peripheral supporting roles and technical support required to make it a success. In addition, it is important that there are identified leaders within company management (champions of the process) who will embrace the philosophy strongly enough to support these teams and technical experts, both in terms of acknowledgments as well as required company resources.

The result, hopefully, is a highly motivated team supported by a company management that will consider and help modify current paradigms and company procedures to make these necessary adjustments to project implementation.

14.4.2 A Different Academic Teaching Approach

Training for the generation of individuals who will make such quantum changes like modularization and standardization must start early. It begins where youth are encouraged to be inquisitive, work for a common goal, be team-oriented, open-minded, willing to try something different, and willing to accept failure as a necessary part of learning. These ideals must be further reinforced in early schooling, where students are encouraged and even challenged to consider alternative solutions while concurrently being taught to understand the basics.

This alternative "out of the box" thinking should become an integral part of their psyche, to the point that by the time they reach high school and college, they are searching for specific educational challenges from whatever fields of interest the students are involved with. In this case, we would hope it would be something to do with modular and standardization philosophy. But such challenges will only take form and continue in an educational environment that combines the standard historic academic teaching with an up-to-date, hands-on approach to learning provided by industry.

Academia must strive to provide more practical applications for the students. This means incorporating more industry experts into the teaching process who have actually worked day-to-day in these sectors. It means bringing the teaching "to life" and making it real.

14.4.3 Identify, Acknowledge, and Incorporate Required Paradigm Shifts

These are critical in so many aspects of our industry on so many levels. An industry either adapts or dies. Change is critical to success. For example, consider a few of the following changes with respect to our industry:

- Stick-built to Modular to Standardized to Modular Standardized Project (MSP)

- Analog to Digital

- Sketches to Drawings to Photos to Point Cloud representations to Digital Twin

- 2D schematics to 3D models to 4D (time) to 5D (cost) to 6D (lifecycle) to NextD?

- Individual databases to Blockchain.

And the list goes on. But, as we have lamented, our particular industry has not kept up with our buddies in manufacturing and other related industries in terms of incorporating true technology breakthroughs. This was mentioned in Chapter 1, where we briefly touched on the automobile, aircraft, and shipbuilding industries and how they reacted to difficulties in their industry and transformed themselves to grow. Our industry must follow a similar transition to survive.

14.4.3.1 Stick-Built to Modular to Standardized

As an example, it should start with a shift in the initial approach to industrial project development. Just like the current concept shift that is occurring from developing a project based on a stick-built philosophy to one that is based on a modular philosophy, the industry now needs to make the next shift from customization as a base case to the use of the standardization planning as the base case. And, only when the avenues for standardization have been exhausted, should a project fall back on the current *modus operandi* of customization for everything, just as only when all avenues for modularization have been exhausted should the project drop back to a stick-built philosophy.

There are corporate constraints with adopting such an approach. It will take a strong management lead from the owner/client initially to make this shift in project philosophy a reality. And, with many EPC firms' profits tied to the simple sale of engineering and other project execution work-hours, this philosophy shift may directly impact profits. So, for this industry shift to truly be incorporated and take hold, the direct tie of profits to work-hours sold must be broken, and the hole created by these lost profits be filled by the benefits derived from creative contracting, profit and risk sharing, liability waivers, etc., to keep these companies whole while advancing our industry toward this next step-change. This brings us to the next accelerator.

14.4.4 Friendly Contracting

Sounds a bit like a paradox or oxymoron. But, as the authors have seen, some of the best examples of successful projects are where extra steps were taken to move away from the traditional "he said/she said" language of a transactional driven contract to a more inclusive sharing of both risk and reward.

CII RT-341 (Construction Industry Institute, 2019) and (Barutha *et al.*, 2021) provide a great pathway on this topic. The research paper by the CII shows a comparison of where our industry currently is (bi-lateral contracts, risk transfer, hierarchical decision-making, win-lose approach, and "best for me" culture) to where the best future exists (relational contracting, multi-party agreements, shared risk/reward, equitable decision-making, lean methods and tools, liabilities waiver, and "best for project" culture)

(Construction Industry Institute, 2019). Through extensive research, they found that the performance certainty of industrial projects is improved by a higher degree of collaboration and integration and demonstrated that industrial projects could also benefit from a more integrated project delivery (IPD) method (Barutha et al., 2021).

Friendly contracting is not just a simple contracting strategy re-vamp. It involves shifting from a "me" world to a "we" world, from win-lose to equitable treatment to all, from risk transfer to risk sharing. This goes beyond a simple contract document, requiring an overall company program shift and from there to an overall industry shift.

14.4.5 Industry Re-branding (or Making the Industry Sexy, Again)

If there is one constant about this industry, it is the fact that it is cyclical and ever-changing. The industry has fallen in and out of favor numerous times, several just in our lifetime. Pick any 5- to 10-year period, and one can find both periods of high craft demands for heavy (industrial) construction and the accompanying increased activity within the oil and chemicals industries followed by periods where very few major capital projects were being pursued, resulting in the mass exodus of craft labor from the industry.

As we move into the future, the concern is that it seems to be getting harder to ramp back up after each downturn. Skilled craft labor (welders, pipefitters, etc.) has been and continues to be in demand and is immediately incorporated into the industry as it becomes available. This is good in terms of providing challenges to those who want to make a career in the industry through a skilled craft. But, this craft has also become more specialized, making it more difficult to replace any experience that leaves through retirement or reduction in force. In addition, the industry seems to be seeing fewer young people actually choosing these crafts as a profession.

So, what can be done?

14.4.5.1 Education and Learning

First, properly support and encourage the education in and learning of these craft skills. We are starting to see renewed efforts, not only from our industry but by the technical crafts themselves, to publicize the benefits of the craft ladder as a profession and a good alternative to the academic university and college route. This should continue to be stressed. These skills have been and will continue to be in high demand. We need to make sure that they are not trivialized by the glamor of the college degree, which for many individuals ends up being no more than trying to fit the expensive square peg of learning into the round hole of a useful application.

14.4.5.2 Innovative Technologies

Second, technology applications supporting these large industry projects need to be emphasized and pushed. Concepts like Blockchain or Digital Twins or Generative Design or Machine Learning are now realities. It takes only a second to see the success in terms of sales in the multitude of gaming platforms developed around the themes of "building" anything and everything and the interest that younger people have in them. Why can't these same successful gaming concepts be combined and incorporated into the industry technology? What if we could make the development of the industrial 3D and 4D models as exciting for the users as these games? We have already seen major plant modeling go from representations of two-dimensional piping drawings as handmade plastic modules, to various levels of complexity in electronic 3D modeling, to where we can now merge the 3D model with the on-site plant via point clouds of dimensional data with an accuracy that allows tie-in spools to be designed and fabricated. So, what about similar step changes in other areas?

Sure, some of this developing technology is currently ongoing in colleges and universities, e.g., see Choi's example efforts here (Choi, 2018; Ghimire et al., 2021; Kithas and Choi, 2022). But, as mentioned in the training enabler, such concept development thinking should be started much earlier, for example, in the elementary, junior high, and high school curriculums. The key to this is simple—make the subject interesting and entertaining to the students and then turn them loose to go and create.

Same thing with industry workers. It can start with a truly engaged workforce (that may or may not be college-trained or prepped) who have the ability to combine BOTH an idea of what is possible with an accurate understanding of how the industry is currently operating and the freedom provided by their management to challenge the status quo.

14.4.5.3 *More Diverse and Inclusive Environments*

As we move into the future, the construction industry needs to put more effort into creating more diverse and inclusive environments. The industry is suffering from labor shortage problems and experiencing challenges in recruiting and retaining skilled craft laborers. In particular, compared to other industries, the construction industry has low interest and participation from underrepresented demographic groups, including women (Choi *et al.*, 2018; Manesh *et al.*, 2020; Shrestha *et al.*, 2020). Modularization can help the industry attract a more diverse workforce. As we explained in Chapter 2, the typical module fabrication shop provides an attractive alternative to the historical site work regimen. Workers are provided with a more standard work week, job security, set benefits, a safe place to work and hone their skills, and less commuting time (resulting in more time with family and a healthier work/lifestyle split of their time).

However, such enticements may still not be enough. The industry needs to create more diverse and inclusive environments, including removing discrimination, inequality, inequity, bias, segregation, and barriers in workplaces (Choi, Shane, and Chih, 2022). Additional information can be found in Dr. Choi's papers on enhancing diversity and inclusion in the engineering-construction industry (Choi *et al.*, 2018; Manesh *et al.*, 2020; Manesh, Choi, and Shrestha, 2020; Shrestha *et al.*, 2020; Choi, Shane, and Chih, 2022).

The potential of combining cutting-edge technology with a craft skill that one can hone via the relatively constant workload provided by the module fabrication facility, provides a great opportunity for the craft worker, regardless of their physical size, ethnicity, or background. These opportunities should be promoted at all levels of education as well as in-house industry training programs.

14.4.6 More Alliances and Research

Alliances between academia and industry must be further identified, developed, and pursued. Like those of years past, the groups from private industry and academic institutions must continue to work together to solve problems greater than both by combining evolving research with existing industry know-how. Academia can provide cutting-edge research but needs direction on where and how such research can be implemented to provide the greatest impact. Industry can identify issues, what is needed, and where it would be best received but may lack the resources to develop research study efforts individually and resolve them. Together, however, they can identify the need and find solutions for industry problems that will be useful and practical enough to be incorporated into the current working mechanisms of the individual corporate company's business model.

While our universities still receive "grants" and other financial support to perform some of this cutting-edge development, they really need to get back to the more historic alliances developed between industry and research or industry and government or designers and application experts, etc. These were historically very successful because such alliances had the dual goal of developing state-of-the-art technologies that were designed to correct or fix a specific industry problem or challenge. The result was an answer to an industry-wide problem that no individual company had the resources to address or fix.

To get this moving, it needs the corporate heads of industry to embrace it. To do so, our industry needs to get back into a position where all companies have what the typical family would have called "disposable income." Income that is available to research, teach, train, try, retry, fail, but eventually succeed on either developing or at least incorporating this new and cutting-edge technology. Our industry needs to get back to the higher level of research efforts they previously supported, both internally and externally.

14.4.7 Planning Techniques and Their Combinations

The aim of this accelerator is to move away from the traditional methods of planning where older tools used are selected based on their historic use in the industry and where newer tools are implemented based on how they are perceived to function without proper practical testing, with both efforts resulting in planning tools that must be forced into a planning situation where they may not fit.

By developing a better understanding of the bases for all the various planning concepts, there is a potential opportunity of creating new or unique planning concepts form-fitted to our industry, perhaps as simple as selecting "pieces" of each of several tools and combining them to meet specific needs.

There are so many variations on the planning theme (Collaborative / Interactive/Effective Interfacing/LEAN/Path of Construction/Path of Engineering Design/Last Planner/ Bag & Tag/Advanced Work Packaging (AWP) and all its derivatives) that trying to select a single specific planning tool or effort that is best for a project could be a mind-boggling effort. And just as the shipping industry ended up developing the unique Interim Product Database concept as an answer to their specific industry issues on how they would mix and match large pieces of pre-designed vessel parts, our industry may also need to examine what the issues are in terms of cost and planning and see if there needs to be a planning technique that might be a unique fit for it. But, in order to develop such a custom-fitted solution, there must be wider knowledge in terms of understanding the bases of all of these concepts at a level where the owner/clients and EPC companies can begin considering a solution that may mix and match parts of them to come up with such a better planning tool.

We will not try to predict what such a composite planning tool might look like for this industry (our crystal ball is not that good)! It may be something completely new, or it may simply introduce a unique addition to an existing planning tool to produce this step change. Who knows? But it cannot move forward until a broader effort is made at understanding what the planning issues are and what is out there as a resource.

14.5 The End or the Beginning?

This is the end of our prepared thoughts. But hopefully, this will be the beginning of yours.

Thanks for the opportunity to allow us to try to influence you.

References

Barutha, P.J. *et al.* (2021) Evaluation of the Impact of Collaboration and Integration on Performance of Industrial Projects. *Journal of Management in Engineering*, 37(4), 04021037. doi: 10.1061/(ASCE)ME.1943-5479.0000921.

Choi, J.O. (2018) A New Graduate Course on Modular Construction: University of Nevada, Las Vegas. In *2018 Modular and Offsite Construction (MOC) Summit*. Hollywood, FL, p. 8.

Choi, J.O. *et al.* (2018) An Investigation of Construction Workforce Inequalities and Biases in the Architecture, Engineering, and Construction (AEC) Industry. In *Construction Research Congress (CRC) 2018*. Reston, VA: American Society of Civil Engineers (ASCE).

Choi, J.O., Shane, J.S. and Chih, Y.Y. (2022) Diversity and Inclusion in the Engineering-Construction Industry. *Journal of Management in Engineering*, 38(2), 02021002. doi: 10.1061/(ASCE)ME.1943-5479.0001005.

Construction Industry Institute (2019) FR-341 – Integrated Project Delivery for Industrial Projects. Austin, TX: The University of Texas at Austin: Construction Industry Institute. Available at: https://www.construction-institute .org/resources/knowledgebase/knowledge-areas/ project-program-management/topics/rt-341/pubs/fr-341.

Ghimire, R. *et al.* (2021) Combined Application of 4D BIM Schedule and an Immersive Virtual Reality on a Modular Project: UNLV Solar Decathlon Case. *International Journal of Industrialized Construction*, 2(1), 1–14. doi: 10.29173/IJIC236.

Kithas, K.A. and Choi, J.O. (2022) Hands-on Education Module for Modular Construction, 3D Design, and 4D Schedule. In *International Conference on Construction Engineering and Project Management (ICCEPM) 2022*. Las Vegas, NV, p. 8.

Manesh, S.N. *et al.* (2020) Spatial Analysis of the Gender Wage Gap in Architecture, Civil Engineering, and Construction Occupations in the United States. *Journal of Management in Engineering*, 36(4), 04020023. doi: 10.1061/(ASCE)ME.1943-5479.0000780.

Manesh, S.N., Choi, J.O., and Shrestha, P. (2020) Critical Literature Review on the Diversity and Inclusion of Women and Ethnic Minorities in Construction and Civil Engineering Industry and Education. In *Construction Research Congress 2020: Safety, Workforce, and Education – Selected Papers from the Construction Research Congress 2020*, pp. 175–184. doi: 10.1061/9780784482872.020.

Shrestha, B.K. *et al.* (2020) Employment and Wage Distribution Investigation in the Construction Industry by Gender. *Journal of Management in Engineering*, 36(4), 06020001. doi: 10.1061/(ASCE)ME.1943-5479.0000778.

15 Key Literature and Resources on Modularization

As a follow-up to the future actions discussed in Chapter 14, this chapter recommends some useful reports, tools, and academic papers on modularization and design standardization—many of which were referenced at one time or the other in the preceding chapters.

15.1 Key Literature from Construction Industry Institute (CII) and Dr. Choi

The Construction Industry Institute (CII), "based at the University of Texas at Austin, is a consortium of more than 140 leading owner, engineering-contractor, and supplier firms from both the public and private arenas" (Construction Industry Institute, 2022). CII has been a leader in improving the construction industry through research and many initiatives. There have been five core research teams focused on modularization in the past. After Research Team (RT-283), Dr. Choi and the Modularization Community for Business Advancement (MCBA) further developed tools and conducted studies on modularization. In 2022, a new research team on modularization (RT-396) was established to develop a new Business Case Analysis Guide for Industrial Modularization. We have briefly introduced or summarized those key literature and resources below.

15.1.1 The Constructability Task Force 83-3

The first research report on modularization was published in 1987 (Tatum, Vanegas and Williams, 1987), which was supported by CII.

Tatum, C.B., Vanegas, J.A., and Williams, J.M. (1987) *Constructability Improvement Using Prefabrication, Preassembly, and Modularization.* Austin, TX: The University of Texas at Austin: Construction Industry Institute.

This report is one of the important and key resources as they established the foundation of the body of knowledge on modularization. They have: (1) provided many key definitions, including prefabrication, pre-assembly, module, and modularization; (2) described the practices of prefabrication, pre-assembly, and modularization in both industrial and building sectors; (3) identified key advantages and barriers through multiple case projects; and (4) provided recommendations for effective implementation of modularization. As they set the first stone, not only the authors of this book but also many succeeding research teams adopted many of their key findings and definitions and referenced them. Several findings from the case projects are outdated as the study was conducted over 35 years ago, but there is still much valuable knowledge that is still useful, such as key definitions.

15.1.2 Research Team 29

The second research team on modularization supported by CII is the following (Construction Industry Institute, 1992):

Construction Industry Institute (1992) *MODEX: Automated Decision Support System for Modular Construction.* Austin, TX: The University of Texas at Austin: Construction Industry Institute.

The research team (RT-29) developed a tool to support the decision on modularization in the early phases of a project. However, this report is not recommended as a reference as the report has been archived and superseded by later publications, which are described below.

15.1.3 Research Team 171

The third research team (RT-171) developed a decision framework and an Excel judgment-based tool that help identify a project's suitability for implementing modularization (Construction Industry Institute, 2002).

Construction Industry Institute (2002) *Implementing the Prefabrication, Preassembly, Modularization, and Offsite Fabrication Decision Framework: Guide and Tool.* Austin, TX: The University of Texas at Austin: Construction Industry Institute.

Later, the academics (Song *et al.*, 2005) from the research team published a journal paper based on this research:

Song, J., Fagerlund, W.R., Haas, C.T., Tatum, C.B., and Vanegas, J.A. (2005) Considering Prework on Industrial Projects. *Journal of Construction Engineering and Management*, 131(6), pp. 723–733.

The tool included a set of questions that, if responded to honestly, would indicate whether a particular project should be modular or conventional stick-built. Also, they were the first ones to introduce the term "PPMOF," which is Prefabrication, Pre-assembly, Modularization, and Offsite Fabrication. The term PPMOF and their tool have been used and loved by the industry for a long time. However, there has been a complaint by the practitioners that the tool, including the factors and the examples in the tool, is outdated as it was published almost 20 years ago.

15.1.4 Research Teams 232 and 255

The fourth (RT-232; Construction Industry Institute, 2007) and fifth (RT-255; Construction Industry Institute, 2011) research teams investigated the shipbuilding industry.

Construction Industry Institute (2007) *Examination of the Shipbuilding Industry.* Austin, TX: The University of Texas at Austin: Construction Industry Institute.

Construction Industry Institute (2011) *Transforming Modular Construction for the Competitive Advantage through the Adaptation of Shipbuilding Production Processes to Construction.* Austin, TX: The University of Texas at Austin: Construction Industry Institute.

Their main recommendations to the construction industry, based on the investigation of the shipbuilding industry, are transitioning from a "stick-built" to a "product-oriented" philosophy by implementing modularization, design standardization, and the interim product database (IPD).

15.1.5 Research Team 283

The newest CII research team was RT-283 (Construction Industry Institute, 2013; O'Connor, O'Brien, and Choi, 2013).

Construction Industry Institute (2013) *Industrial Modularization: Five Solution Elements.* Austin, TX: The University of Texas at Austin: Construction Industry Institute.

O'Connor, J.T., O'Brien, W.J., and Choi, J.O. (2013) *Industrial Modularization: How to Optimize; How to Maximize.* Austin, TX: The University of Texas at Austin: Construction Industry Institute.

Beyond RT-171 research, RT-283 developed 5 Solution Elements critical to modularization:

1. Business Case Process
2. Execution Plan Differences
3. Critical Success Factors
4. Standardization Strategy
5. Maximization Enablers.

As we mentioned in the previous chapters, the authors of this book were part of this research team, thus, we might be biased! But we believe the resulting reports and conclusions from CII RT-283 are the most useful in terms of up-to-date knowledge on industrial modularization. That is why we adopted and borrowed some knowledge and key learnings from these reports for inclusion in this book:

- Chapter 5 (The Business Case for Modularization) shows the influences of Solution Element #1. Business Case Process.

- Chapter 6 (Universal Modularization Requirements) shows the influences of Solution Element #2. Execution Plan Differences.

- Chapter 7 (Key Critical Success Factors for Modular Project Success) shows the influences of Solution Element #3. Critical Success Factors.

- Chapter 11 (Standardization) shows the influences of Solution Element #4. Standardization Strategy.

- Chapter 13 (Moving Forward: Future Actions) shows the influences of Solution Element #5. Maximization Enablers.

For practitioners, we highly recommend the reports from RT-283, which are available at <https://www.construction-institute.org/resources/knowledgebase/best-practices/planning-for-modularization/topics/rt-283 >

Later, the academics, including one of the book authors, published multiple academic papers (Choi and O'Connor, 2014, 2015; O'Connor, O'Brien, and Choi, 2014, 2015, 2016, 2017; O'Brien, O'Connor, and Choi, 2015; Choi et al., 2019).

Choi, J.O., and O'Connor, J.T. (2014) Modularization Critical Success Factors Accomplishment: Learning from Case Studies. In *Construction Research Congress 2014*, pp. 1636–1645. https://doi.org/10.1061/9780784413517.167.

Choi, J.O., and O'Connor, J.T. (2015) Modularization Business Case Analysis: Learning from Industry Practices Tool. In M. Al-Hussein, O. Moselhi, S. Kim, and R.E. Smith (Eds.), *2015 Modular and Offsite Construction (MOC) Summit and 1st International Conference on the Industrialization of Construction (ICIC)*. Edmonton, Alberta: Dept. of Civil and Environmental Engineering. University of Alberta, pp. 69–76.

Choi, J.O., O'Connor, J.T., Kwak, Y.H., and Shrestha, B.K. (2019) Modularization Business Case Analysis Model for Industrial Projects. *Journal of Management in Engineering*, 35(3), 04019004. https://doi.org/10.1061/ (ASCE) ME.1943-5479.0000683.

O'Brien, W.J., O'Connor, J.T., and Choi, J.O. (2015) Modularization Business Case: Process Flowchart and Major Considerations. In M. Al-Hussein, O. Moselhi, S. Kim, and R.E. Smith (Eds.), *2015 Modular and Offsite Construction (MOC)*

Summit and 1st International Conference on the Industrialization of Construction (ICIC), pp. 60–67. Edmonton, Alberta: Dept. of Civil and Environmental Engineering. University of Alberta.

O'Connor, J.T., O'Brien, W.J., and Choi, J.O. (2014) Critical Success Factors and Enablers for Optimum and Maximum Industrial Modularization. *Journal of Construction Engineering and Management*, 140(6), 04014012. https://doi.org/10.1061/ (ASCE)CO.1943-7862.0000842.

O'Connor, J.T., O'Brien, W.J., and Choi, J.O. (2015) Standardization Strategy for Modular Industrial Plants. *Journal of Construction Engineering and Management*, 141(9), 4015026. https://doi.org/10.1061/(ASCE)CO.1943-7862.0001001.

O'Connor, J.T., O'Brien, W.J., and Choi, J.O. (2016) Industrial Project Execution Planning: Modularization versus Stick-Built. *Practice Periodical on Structural Design and Construction*, 21(1), 04015014. https://doi.org/10.1061/(ASCE) SC.1943-5576.0000270.

O'Connor, J.T., O'Brien, W.J., and Choi, J.O. (2017) Industry-Wide Maximization Enablers for Higher Levels of Modularization. In Hosin "David" Lee (Ed.), *The First International Conference on Maintenance and Rehabilitation of Constructed Infrastructure Facilities (MAIREINFRA1)*, pp. 1–6. International Society for Maintenance and Rehabilitation of Transport Infrastructures.

These academic journal papers and conference papers cover in-depth literature reviews, research methodology, improved models and frameworks, detailed findings, and discussions beyond what was presented in the reports. For those interested in the details, refer to the above papers. These papers will complement the book.

15.1.6 Dr. Choi's Dissertation and the Follow-up Research

After CII RT-283 research, there was a need from the industry to better understand the relative significance of modularization critical success factors (CSFs) identified by RT-283 and their association with performances of a modular project. To address this need, Dr. Choi identified the correlations between the CSFs' accomplishments and project performance (Choi, 2014).

Choi, J.O. (2014) Links between Modularization Critical Success Factors and Project Performance. Ph.D. dissertation. The University of Texas at Austin.

After his dissertation research, Dr. Choi, with his colleagues, further identified the recipes for cost and schedule success in modular projects (Choi, O'Connor, and Kim, 2016) and calibrated CII RT-283's recommendation timings for CSFs (Choi et al., 2019).

Choi, J.O., O'Connor, J.T., and Kim, T.W. (2016) Recipes for Cost and Schedule Successes in Industrial Modular Projects: Qualitative Comparative Analysis. *Journal of Construction Engineering and Management*, 142(10), 04016055. https://doi .org/10.1061/(ASCE)CO.1943-7862.0001171.

Choi, J.O., O'Connor, J.T., Kwak, Y.H., and Ghimire, R. (2019) Calibrating CII RT-283's Modularization Critical Success Factor Accomplishments. *Modular and Offsite Construction (MOC) Summit Proceedings*, pp. 235–242. https://doi .org/10.29173/MOCS99.

15.1.7 CII Modularization Community for Business Advancement (MCBA) Toolkit

The Modularization Community for Business Advancement (MCBA) is a formal venue for the exchange of knowledge that is useful in planning, designing, and executing modularization of varying complexity on capital facility projects (Construction Industry Institute, 2022a). The authors have been supporting this group since 2013 as chairperson or academic advisor. In 2015, MCBA, led by the authors, developed a toolkit that consists of four tools:

- Tool #1: Modularization Business Case Analysis Tool (.xlsx)
- Tool #2: Modular Project Execution Planning Guide (.pdf)
- Tool #3A: Intro to Modularization – Why Modularize (.ppt)
- Tool #3B: Intro to Modularization – How to Optimize (RT-283 Overview) (.ppt)
- Tool #4: Critical Success Factor (CSF) Accomplishment Tracking Tool (.xlsx).

Their goal was to develop some practical applications of the research that the industry would be able to pick up and use. Tool #1 is an Excel-based tool that helps practitioners conduct module economics analysis; Tool #2, which is intended as a handy reference, identifies key activities and considerations by project phase; Tool #3 summarizes and

navigates RT-283 reports; and Tool #4 is an Excel score sheet to benchmark application of CFS. Tool #1 was explained in detail in Chapter 5. If your company is a CII member, you can download the modularization toolkit from here: https://www.construction-institute.org/resources/ knowledgebase/knowledge-areas/modularization/ topics/rt-283.

15.1.8 (Ongoing) Research Team 396

Most recently, in 2022, CII launched a new research team (RT-396) to develop a new Business Case Analysis Guide for Industrial Modularization. This new research team is expected to develop not only: (1) a new tool that can determine whether a project should modularize definitively and determine the optimal extent of modularization for a given project considering new business drivers, but also (2) an Advanced Modularization Planning Guide (Construction Industry Institute, 2022b). The authors are also participating in this research team as Primary Investigator (PI) and lead. Readers are recommended to check the CII website for future publications from this research team.

15.2 Key Research Articles on Modularization and Standardization by Dr. Choi

Dr. Choi has published over 30 publications on modularization and standardization in the past 10 years, not only in the industrial sector but also in the building and civil sectors. Other than the publications listed above, we share the rest of the key papers for those who are interested by topic. The focus topics include:

1. New technologies and approaches for modularization (Lee et al., n.d.; Choi and Kim, 2019; Choi, Shrestha, Kwak, et al., 2020b; Khodabandelu et al., 2020; Ghimire et al., 2021; Lee, Choi, and Song, 2022; Song, Choi, and Lee, 2022).

2. Modular construction in the building sector (Choi, Chen, and Kim, 2019; Paliwal et al., 2021; Prestia et al., 2021; Choi, Lee, and Weber, 2022).

3. Modular construction for post-disaster reconstruction (Ghannad, Lee, and Choi, 2019, 2020, 2021; Colletta, Lim, and Choi, 2021; Harris et al., 2022).

4. Education for modularization (Choi, 2018; Kithas and Choi, 2022).

5. Accelerated bridge construction (Sakhakarmi, Choi, and Park, 2018; Prajapati and Choi, 2019).

6. Facility standardization (Choi, Shane, *et al.*, 2018; Construction Industry Institute, 2019a, 2020; Choi, Shrestha, Kwak, *et al.*, 2020a; Choi, Shrestha, Shane, *et al.*, 2020; Shrestha, Choi, Kwak, *et al.*, 2020a; Shrestha, Choi, Kwak, *et al.*, 2020b; Choi, Shrestha, Kwak *et al.*, 2021a; 2021b; Shrestha *et al.*, 2021; Choi *et al.*, 2022).

The full list of publications by topic can be found below.

15.2.1 New Technologies and Approaches for Modularization

Choi, J.O., and Kim, D.B. (2019) A New UAV-Based Module Lifting and Transporting Method: Advantages and Challenges. In *Proceedings of the 36th International Symposium on Automation and Robotics in Construction, ISARC 2019*, pp. 645–650. https://doi.org/10.22260/isarc2019/0086.

Choi, J.O., Kwak, Y.H., Shane, J.S., and Shrestha, B.K. (2019) Identifying Potential Innovative Technologies and Management Approaches for Design Standardization. In *Computing in Civil Engineering 2019: Visualization, Information Modeling, and Simulation – Selected Papers from the ASCE International Conference on Computing in Civil Engineering 2019*, pp. 256–263. https://doi.org/10.1061/9780784482421.033.

Choi, J.O., Shrestha, B.K., Kwak, Y.H., and Shane, J.S. (2020) Innovative Technologies and Management Approaches for Facility Design Standardization and Modularization of Capital Projects. *Journal of Management in Engineering*, 36(5), 04020042. https://doi.org/10.1061/(ASCE)ME.1943-5479.0000805.

Ghimire, R., Lee, S., Choi, J.O., Lee, J.-Y., and Lee, Y.-C. (2021) Combined Application of 4D BIM Schedule and an Immersive Virtual Reality on a Modular Project: UNLV Solar Decathlon Case. *International Journal of Industrialized Construction*, 2(1), 1–14. https://doi.org/10.29173/IJIC236.

Khodabandelu, A., Choi, J.O., Park, J., and Sanei, M. (2020) Developing a Simulation Model for Lifting a Modular House. In *Construction Research Congress 2020: Computer Applications: Selected Papers from the Construction Research Congress 2020*, pp. 145–152. https://doi.org/10.1061/9780784482865.016.

Lee, D., Wen, L., Choi, J., and Lee, S. (2023) Sensor Integrated Hybrid Blockchain System for Supply Chain Coordination in Volumetric Modular Construction. *ASCE Journal of Construction Engineering and Management,* 149(1). https://doi.org/10.1061/(ASCE)CO.1943-7862.0002427.

Lee, S., Choi, J.O., and Song, S. (2022) Cutting-Edge Technologies to Achieve a Higher Level of Modular Construction: Literature Review. In *Proceedings of International Conference on Construction Engineering and Project Management (ICCEPM) 2022*, pp. 536–542.

Song, S., Choi, J.O., and Lee, S. (2022) The Current State and Future Directions of Industrial Robotic Arms in Modular Construction. In *Proceedings of International Conference on Construction Engineering and Project Management (ICCEPM) 2022*, pp. 336–343.

15.2.2 Modular Construction in the Building Sector

Choi, J.O., Chen, X.B., and Kim, T.W. (2019) Opportunities and Challenges of Modular Methods in Dense Urban Environment. *International Journal of Construction Management*, 19(2), pp. 93–105. https://doi.org/10.1080/15623599.2017.1382093.

Choi, J.O., Lee, S., and Weber, E. (2022) Lessons Learned during the Early Phases of a Modular Project: A Case Study of UNLV's Solar Decathlon 2020 Project. In *Proceedings of International Conference on Construction Engineering and Project Management (ICCEPM) 2022*, pp. 543–550.

Paliwal, S., Choi, J.O., Bristow, J., Chatfield, H.K., and Lee, S. (2021) Construction Stakeholders' Perceived Benefits and Barriers for Environment-Friendly Modular Construction in a Hospitality Centric Environment. *International Journal of Industrialized Construction*, 2(1), pp.15–29. https://doi.org/10.29173/IJIC252.

Prestia, J., Choi, J.O., Lee, S., and James, D. (2021) Lessons Learned from UNLV's Solar Decathlon 2017 Competition Experience: Design and Construction of a Modular House. In *Proceedings of the Canadian Society of Civil Engineering Annual Conference 2021*, pp. 565–575.

15.2.3 Modular Construction for Post-Disaster Reconstruction

Colletta, A., Lim, J., and Choi, J.O. (2021) Review of Infrastructure Resiliency Policy for Natural Disasters. In *Proceedings of the Canadian Society of Civil Engineering Annual Conference 2021*, pp. 637–648.

Ghannad, P., Lee, Y.-C., and Choi, J.O. (2019) Investigating Stakeholders' Perceptions of Feasibility and Implications of Modular Construction-Based Post-Disaster Reconstruction. In *Modular and Offsite Construction (MOC) Summit Proceedings*, pp. 504–513. https://doi.org/10.29173/MOCS132.

Ghannad, P., Lee, Y.-C., and Choi, J.O. (2020) Feasibility and Implications of the Modular Construction Approach for Rapid Post-Disaster Recovery. *International Journal of Industrialized Construction*, 1(1), pp. 64–75. https://doi.org/10.29173/IJIC220.

Ghannad, P., Lee, Y.-C., and Choi, J.O. (2021) Prioritizing Postdisaster Recovery of Transportation Infrastructure Systems Using Multiagent Reinforcement Learning. *Journal of Management in Engineering*, 37(1), 04020100. https://doi.org/10.1061/(ASCE)ME.1943-5479.0000868.

Harris, W., Choi, J.O., Lim, J., and Lee, Y.-C. (2022) Long-Term Wildfire Reconstruction: In Need of Focused and Dedicated Pre-Planning Efforts. In *Proceedings of International Conference on Construction Engineering and Project Management (ICCEPM) 2022*, pp. 923–928.

15.2.4 Education for Modularization

Choi, J.O. (2018) A New Graduate Course on Modular Construction: University of Nevada, Las Vegas. In *Proceedings of the 2018 Modular and Offsite Construction Summit*, pp. 125–132. https://doi.org/10.29173/mocs48.

Kithas, K.A., and Choi, J.O. (2022) LEGO Education Module for Modular Construction and 4D Scheduling. In *Proceedings of International Conference on Construction Engineering and Project Management (ICCEPM) 2022*, pp. 484–491.

15.2.5 Accelerated Bridge Construction

Prajapati, E., and Choi, J.O. (2019) A Pilot Study of Identifying Execution Plan Differences for Accelerated Bridge Construction. In *Modular and Offsite Construction (MOC) Summit Proceedings*, pp. 198–205. https://doi.org/10.29173/MOCS94.

Sakhakarmi, S., Choi, J.O., and Park, J. (2018) Business Case Process for Accelerated Bridge Construction. In *Proceedings of International Road Federation Global R2T Conference and Expo*, pp. 705–711.

15.2.6 Facility Standardization

Choi, J.O., Shane, J.S., Kwak, Y., and Shrestha, B. (2018) Achieving Higher Levels of Facility Design Standardization in the Upstream, Midstream, and Mining Commodity Sector: Barriers and Challenges. In *Proceedings of Construction Research Congress (CRC) 2018*, pp. 278–287.

Choi, J.O., Shrestha, B.K., Kwak, Y.H., and Shane, J.S. (2020) Critical Success Factors and Enablers for Facility Design Standardization of Capital Projects. *Journal of Management in Engineering*, 36(5), 04020048. https://doi.org/10.1061/(ASCE)ME.1943-5479.0000788.

Choi, J.O., Shrestha, B.K., Kwak, Y.H., and Shane, J. (2021a) Exploring the Benefits and Trade-Offs of Design Standardization in Capital Projects. *Engineering, Construction and Architectural Management*, 29(3), pp. 1169–1193. https://doi.org/10.1108/ECAM-08-2020-0661.

Choi, J.O., Shrestha, B.K., Kwak, Y.H., and Shane, J. (2021b) Facility Design Standardization Work Process and Optimization in Capital Projects. In *Proceedings of the Canadian Society of Civil Engineering Annual Conference 2021*, pp. 491–503.

Choi, J.O., Shrestha, B.K., Shane, J.S., and Kwak, Y.H. (2020) Facility Design Standardization Decision-Making Model for Industrial Facilities and Capital Projects. *Journal of Management in Engineering*, 36(6), 04020077. https://doi.org/10.1061/(asce)me.1943-5479.0000842.

Choi, J.O, Shrestha, B.K., Song, S., Shane, J.S, and Kwak, Y.H. (2022) Facility Design Standardization: Six Solution Pieces and Industry Maximization Enablers. In *Proceedings of Construction Research Congress (CRC) 2022*, pp. 715–723.

Construction Industry Institute (2019) *Achieving Higher Levels of Facility Standardization in the Upstream, Midstream, and Mining (UMM) Commodity Market*. Volume 1: *Four Solution Pieces*. https://www.construction-institute.org/resources/knowledgebase/knowledge-areas/project-program-management/topics/rt-umm-01.

Construction Industry Institute (2020) *Achieving Higher Levels of Facility Standardization in the Upstream, Midstream, and Mining (UMM) Commodity Market*. Volume 2: *Business Case Analysis Model and Work Process*. https://www.construction-institute.org/resources/knowledgebase/knowledge-areas/project-program-management/topics/rt-umm-01.

Shrestha, B.K., Choi, J.O., Kwak, Y.H., and Shane, J.S. (2020a) Timings of Accomplishments for Facility Design Standardization Critical Success Factors in Capital Projects. In *Construction Research Congress 2020: Project Management and Controls, Materials, and Contracts: Selected Papers from the Construction Research Congress 2020*, pp. 898–906. https://doi.org/10.1061/9780784482889.095.

Shrestha, B.K., Choi, J.O., Kwak, Y.H., and Shane, J.S. (2020b) How Design Standardization CSFs Can Impact Project Performance of Capital Projects. *Journal of Management in Engineering*, 36(4), 06020003. https://doi.org/10.1061/(ASCE)ME.1943-5479.0000792.

Shrestha, B.K., Choi, J.O., Kwak, Y.H., and Shane, J.S. (2021) Recipes for Standardized Capital Projects' Performance Success. *Journal of Management in Engineering*, 37(4), 04021029. https://doi.org/10.1061/(ASCE)ME.1943-5479.0000926.

ABBREVIATIONS

Abbreviations and Acronyms	Full Form
3D	three-dimensional
ACCE	Aspen Capital Cost Estimator
AI	artificial intelligence
AIWR	all-in wage rate
ANSI	American National Standards Institute
API	American Petroleum Institute
AR	augmented reality
AWP	advanced work packaging
BD	business development
BFD	block flow diagram
BIM	building information modeling
BOP	blowout preventer
CII	construction industry institute
CMT	construction management team
COG/C of G	center of gravity
Comm	commissioning
connex box	container (express) box
COOEC	China Offshore Oil Engineering Co., Ltd.
CSF	critical success factor
CWA	construction work area
CWP	construction work package
d	day
Demob	demobilization
E&I&C	electrical and instrumentation and controls
Engr	engineering
EPC	engineering, procurement, and construction
EPCM	engineering, procurement and construction management
EPD	execution plan differences
EPR	economic productivity ratio
eval	evaluation
Fab	fabrication
FEED	front end engineering design
FEL	front-end loading
FID	financial investment decision

Abbreviations and Acronyms	Full Form
FWP	fabrication work package
H x L x W	height x length x width
HH	heavy haul
HHI	Hyundai Heavy Industries
HL	heavy lift
hrs	hours
HSE/HS&E	health, safety, and environment
HSSE	health, safety, security and environment
IFC	issued for construction
IPA	independent project analysis
IPD	interim product database
IPD	integrated project delivery
ISO	International Organization for Standardization
IT	information technology
ITB	invitation to bid
IVR	immersive virtual reality
IWP	installation work package
Lic.	licensor
LiDAR	light detection and ranging
LL	long lead
LNG	liquefied natural gas
LO-LO	lift-on-lift-off
LTA	lost time accident
Ltd.	limited company
LTI	lost time incidents
MBI	Modular Building Institute
MCBA	Modularization Community for Business Advancement
MCC	motor control center
Mob	mobilization
MOD	modularization
MOF	module offloading facility
MR	mixed reality
MSP	modular standardized plant
MTL	material
MTO	material take-off
MTPA	million tons per annum

Abbreviations and Acronyms	Full Form
NDE	non-destructive examination
NPV	net present value
O&M	operations and maintenance
OEMs	original equipment manufacturers
OFE	owner furnished equipment
P/R	pipe rack
PAR	pre-assembled rack
PAS	pre-assembled structure
PAU	pre-assembled units
PF	productivity factor
PFD	process flow diagram
PFI	Production Facilities Incorporated
PMT	project management team
PO	purchase order
PPE	personal protective equipment
PPMOF	prefabrication, preassembly, modularization, and off-site fabrication
Pre-comm	pre-commissioning
Prefab	prefabrication
Prep	preparation
QA/QC	quality assurance/quality control
RFID	radio frequency identification
RFQ	request for quote
RFSU	ready for startup
RIR	recordable incident rate
RO	reverse osmosis

Abbreviations and Acronyms	Full Form
ROR	rate of return
RO-RO	roll-on-roll-off
ROW	right of way
RT	research team
RTU	remote terminal unit
RWC	relative work-hour cost
SHI	Samsung Heavy Industries
SIIR	serious injury incident rate
SMU	skid mounted unit
SOI	solicitation of interest
SOW	scope of work
SPMT	self-propelled modular transporter
SRU	sulfur recovery unit
SWHU	single weld hook-up
TIC	total installed cost
TJ	terajoule
TLA	three-letter acronyms
TOOH	trip out of hole
Tran	transportation
UMM	upstream, midstream, and mining
USGC	US Gulf Coast
VAU	vendor assembled unit
VPU	vendor package unit
VR	virtual reality
WH	work-hour
w/r/t	with regard to
XR	extended reality

CONTRIBUTORS

Seungtaek Lee, Ph.D., postdoctoral scholar, Department of Civil and Environmental Engineering and Construction, University of Nevada, Las Vegas, 4505 S. Maryland Parkway, Las Vegas, NV 89154, USA.

Binit Shrestha, Ph.D., Department of Civil and Environmental Engineering and Construction, University of Nevada, Las Vegas, 4505 S. Maryland Parkway, Las Vegas, NV 89154, USA.

INDEX